# JOURNEYMAN EXAM WORKBOOK

**By Tom Henry**

**Based on the 1999 National Electrical Code®**

This workbook contains twelve closed book exams and fouteen open book exams for a total of over **1300** electrical exam questions with answers and references. This workbook is designed to help prepare the electrician for the Journeyman electrical examination.

Fifth printing August 2000

**ISBN 0 - 945495 - 60 - 9**

*ENRY PUBLICATIONS SINCE 1985*

# THE JOURNEYMAN ELECTRICAL EXAMINATION

You must become familiar with the type of exam you are going to take. Make sure you know what subjects are covered on the exam. Find out all you can, how many questions will be asked? What type of questions, multiple-choice, true-false, fill in the blank? Show your work or computer graded? What is the time allowed for the exam? What score is required to pass? These are some of the questions that will help you in your preparation. Get a blueprint of the exam you are to take.

Exams are based on time. You may have one hour to answer 50 closed book questions or two hours to answer 50 open book questions or three hours to answer 30 calculations.

The following is a typical exam format used today.

The exam is given in three parts:

Part I closed book containing 50 questions with a one hour time limit.

Part II open book containing 50 questions with a two hour time limit.

Part III open book containing 30 calculation type questions with a three hour time limit.

This Journeyman exam workbook is designed to prepare the electrician for Part I and Part II of the exam.

**Part I Closed book One hour**
**50 questions from the areas shown below.**

**DEFINITIONS**
**CODE**
**GENERAL KNOWLEDGE**
**BASIC THEORY**
**SAFETY**
**RECOGNITION of TOOLS - EQUIPMENT**

**Part II Open book Two hours**
**50 questions from the areas shown below.**

**CODE**
**GENERAL KNOWLEDGE**
**THEORY**

**CALCULATIONS FOR THE ELECTRICAL EXAM**

Based on the 1999 Code

By Tom Henry

Part III preparation requires the book "Calculations for the Electrical Exam" by Tom Henry.

# PREPARING FOR A CLOSED BOOK EXAM

**Part I Closed Book**

This part of the test is where common sense, apprenticeship and years on the job are helpful. Safety type questions are asked, questions on practical knowledge as the proper connections to a switch circuit. Ohms law and basic theory questions are asked in Part I closed book. If they want to make the closed book exam more difficult, they ask questions from the Code book. Definitions from Article 100 are a favorite closed book question as they expect the electrician to know the definitions. Prior to the exam the last thing the applicant should do is scan Article 100 in the Code. Try to retain as many definitions as you can. The first part of the exam that you are handed is Part I which contains definitions closed book.

A time limit of one hour is allowed to answer the 50 closed book questions. The 50 questions can be answered easily in less than an hour. It's very simple, either you know the answer or you don't. It doesn't help to sit and scratch your head pondering over the correct answer. It has been proven in test taking that the longer you hesitate in selecting the choice, the more likely you are to talk yourself out of the correct answer.

Read the question and the choice of answers **carefully** and select your choice and move to the next question.

After applying the work required by this book you will be able to answer the 50 questions in twenty to thirty minutes. When Part I is completed raise your hand and ask for Part II of the exam. This will provide you with extra time for Part II open book which, I feel, is the most difficult part of the exam.

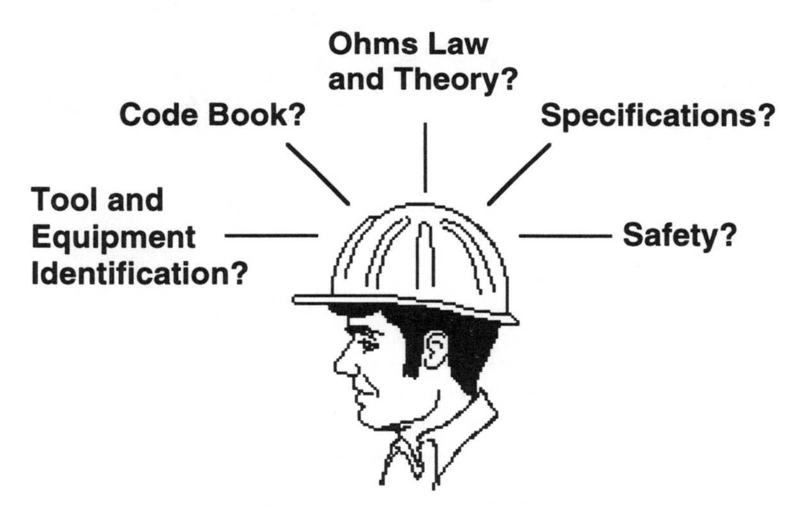


**How much do you know, or can remember of the subjects asked?**

*Excellent study aids to prepare for the closed book exam are the audio tapes Item #192 and the playing cards Item #392.*

## JOURNEYMAN CLOSED BOOK EXAM 50 QUESTIONS

Shown below is how the 50 questions are divided up.

- 20-22 questions are asked from general theory which represents 42% of the exam
- 7-9 questions are asked from field application which represents 16% of the exam
- 4-6 questions are asked from Chapter 1 which represents 10% of the exam
- 4-6 questions are asked from Chapter 2 which represents 10% of the exam
- 4-6 questions are asked from Chapter 3 which represents 10% of the exam
- 1-3 questions are asked from construction specifications which represents 4% of the exam
- 1-3 questions are asked from Chapter 9 which represents 4% of the exam
- 0-2 questions are asked from Chapter 4 which represents 2% of the exam
- 0-2 questions are asked from Article 90 Introduction which represents 2% of the exam

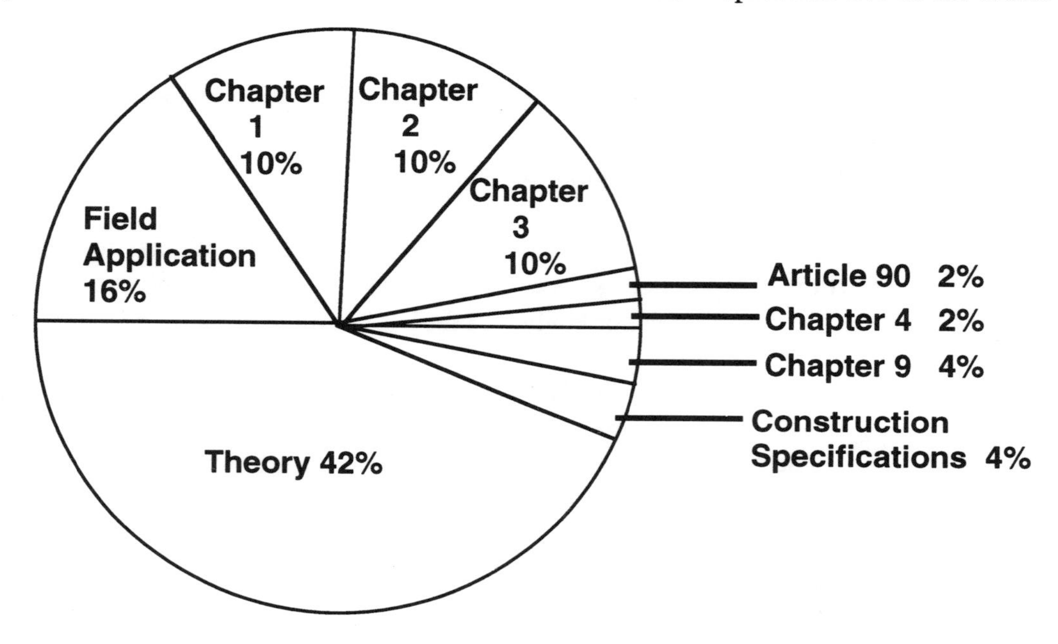


Theory is a big part of an electrical exam. The Ohms Law book and Theory book by Tom Henry are a must!

# PREPARING FOR AN OPEN BOOK EXAM

**Part II Open Book**

Most applicants agree this is the most difficult part of an electrical exam. Time becomes such an important factor. 50 open book questions are to be answered in two hours on the Journeyman exam.

Part II is a test of your knowledge and use of the National Electrical Code. 86% of the open book Journeyman questions are from the Code book.

Your score on the open book exam depends on how familiar you are with the Code book. Most exam applicants run out of time and are not able to find all the answers to the questions within the limited time.

**JOURNEYMAN EXAM**
**50 QUESTIONS**
**TWO HOUR TIME LIMIT**

The key to an open book exam is not to spend too much time on one question. If the question does not contain a key word that you can find in the index, **skip this question**, and continue to the next question. If you spend 3 minutes, 5 minutes, 6 minutes on a question and never find the answer you are eating into the time that should be used for the answers you can find.

In general there are usually 8 to 10 really difficult questions on an exam. The remaining questions after proper preparation you will be able to find within the alotted time. Skip these 8 or 10 as you recognize them and move on finding the other answers. If you answer 40 questions correctly out of a total of 50 questions your score would be 80%! That's better than in some cases where the applicant hasn't even answered 20 questions and time has ran out. You **can't** spend 5 or 6 minutes on a question. Never leave a question unanswered, unanswered is counted wrong. Always select a multiple choice answer before time runs out.

Proper preparation is so important in passing an open book exam. Don't be guilty of reading a question and feeling, "I know the answer so I won't bother looking in the Code book". The following pages will prove how this can be a big mistake. I teach by being properly prepared with how to find your way around in the Code book. You'll be able to look up all the answers within the time limit.

The difficulty occurs when you say Code book. Most applicants taking an exam are not familiar enough with the Code book and it's easy to understand why only 25 out of 100 pass an electrical exam.

## JOURNEYMAN OPEN BOOK EXAM 50 QUESTIONS

Shown below are how the 50 questions are divided up.

•14-16 questions are asked from Chapter 2 which represents 30% of the exam
•14-16 questions are asked from Chapter 3 which represents 30% of the exam
•9-11 questions are asked from Chapter 4 which represents 20% of the exam
•4-6 questions are asked from theory which represents 10% of the exam
•1-3 questions are asked on tools which represents 4% of the exam
•1-3 questions are asked from Articles 100 and 110 which represents 4% of the exam
•0-2 questions are asked from Articles 600 and 680 which represents 2% of the exam

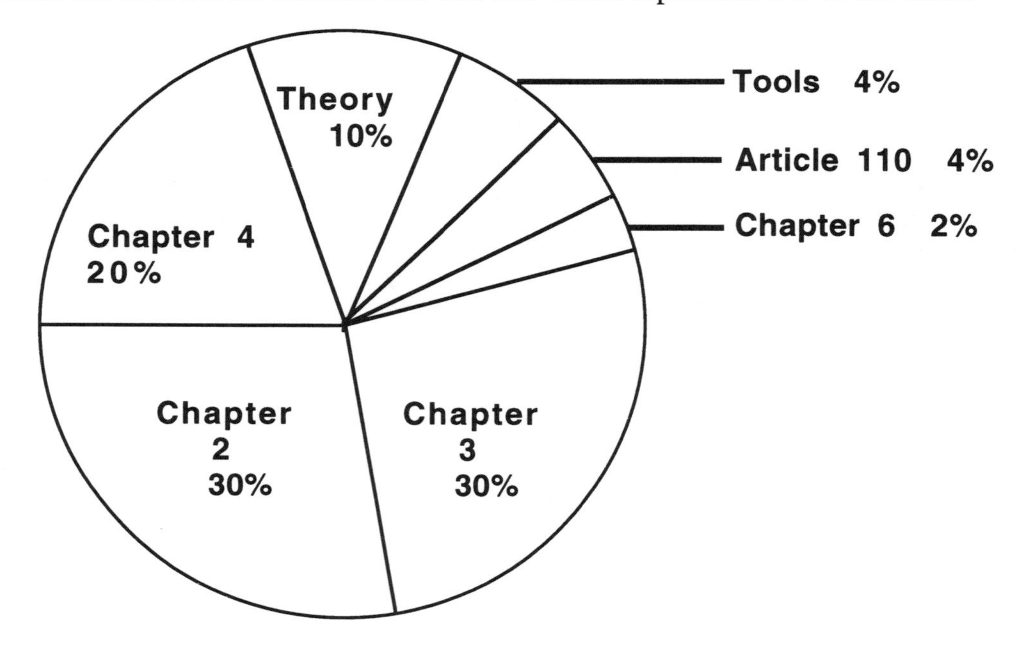

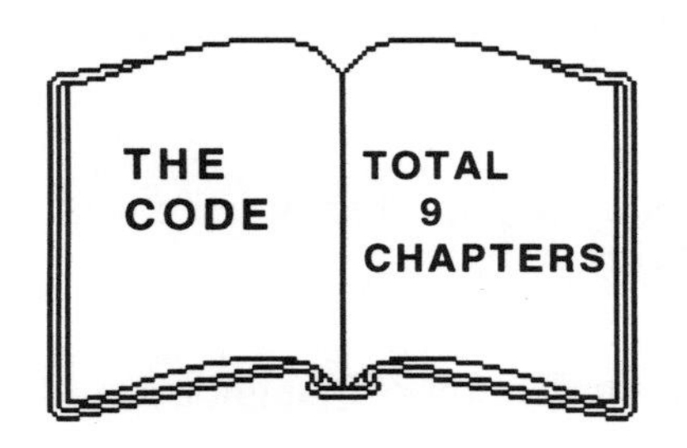


The most difficult task in preparing for the electrical exam is trying to "study" the Code.

The Code book is divided into nine chapters and then divided into articles, parts and sections.

The "meat" of the Code is the first four chapters. General wiring, grounding, services, motors, etc.

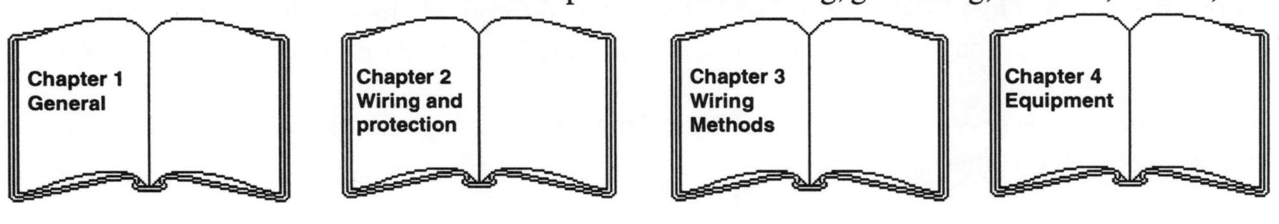


Chapters 5 through 9 are for special applications.

The following is an example of how the Code is divided: Flexible cords are not permitted as a substitute for fixed wiring of a structure per 400-8 of the Code.

The 400-8 is broken down to the 4 indicates Chapter 4.

The 400 is Article 400.

Article 400 is divided into three parts: Part A. General, Part B. Construction Specifications, Part C. Portable Cables over 600 Volts, Nominal.

400-8, the 8 is the section number.

Chapter 4, Article 400, Section 8 which is located in Part A of Article 400.

The latter part of each Article will contain the **over 600 volts (high-voltage) section**.

Example: The definition of a fuse is located in the **over 600 volts** Part B of Definitions Article 100 Chapter 1. Article 100 Definitions is listed in alphabetical order but fuse is not listed in Part A. Following the last Definition in Part A is the word wet location. Part B over 600 volts starts after wet location. Fuse is defined in Part B of Definitions not Part A.

It is very helpful as we try to master the Code book to know how it is laid out in Chapters, Articles, Parts and Sections.

After completing these practice exams turn to the answer sheets in the back of this book and grade yourself. 75% is passing.

To find your percentage simply divide the number of correct answers by the number of questions. Example: 38 correct answers divided by 50 questions would equal 76%.

I teach in my Electrical Code Classes that the key to the exam is that the student must first understand the question, which requires **careful reading of each word.**

As practice for the Open Book Exam Part II, try to find the correct answers for the following seven questions on the next page. Time limit for the 7 questions: 16 to 20 minutes. **GOOD LUCK!!**

After answering the seven open book questions turn to the next page for the answers with references and the index key format.

I chose seven of the more difficult questions from the Journeyman exam so the student will realize how important **TIME** is in this part of the exam. Again, the key is don't spend too much time on a difficult to locate question, **skip** over it and move on, it won't fail you!

This workbook was designed to help you with this difficult area of the exam. Some students purchase a Code book just prior to taking an exam, and as you can see after completing this workbook, you are expected to be an expert in finding references in the Code book. How do you study a Code book for an exam? This workbook is the best way as it **forces** you to find the answers in the time limit. Write down your score and time spent on each exam in this workbook and notice your improvement as you work the latter exams.

To pass the exam is very simple, it takes work! Like with anything in life, you get out of it what you put into it. The more time you spend preparing for the exam the easier it will become.

As you work these exams and grade yourself, hi-lite the answers with a marking pen in your Code book, or better yet purchase the "Ultimate Code Book" which has the complete package for taking an electrical exam.

The Ultimate
Item #518

Will Rogers once said "you can't come back from someplace you've never been". This book will take you there.

## OPEN BOOK PRACTICE 7 QUESTIONS TIME LIMIT: 16 MINUTES

1. Two wire attachment plugs ____.

**(a) need not have their terminals marked for identification**
**(b) need not have their terminals marked for identification unless polarized**
**(c) never need to have their terminals identified**
**(d) always need to have their terminals identified**

2. In a residence, no point along the floor line in any wall space may be more than ____ feet from an outlet.

**(a) 6 (b) 6 1/2 (c) 12 (d) 10**

3. Which of the following statements about a #2 THHN cu conductor is correct?

**(a) Its maximum operating temperature is 90° C**
**(b) It has a nylon insulation**
**(c) Its area is .067 square inches**
**(d) It has a DC resistance of .319 ohms per m/ft from Table 8**

4. The maximum permissible open circuit voltage of electric-discharge lighting equipment used in a dwelling occupancy is ____ volts.

**(a) 1000 (b) 120 (c) 240 (d) 50**

5. Voltage shall not exceed 600 volts between conductors on branch circuits supplying only ballasts for electric-discharge lamps in tunnels with a height of not less than ____ feet.

**(a) 12 (b) 15 (c) 18 (d) 22**

6. Noninsulated busbars will have a minimum space of ____ inches between the bottom of enclosure and busbar.

**(a) 6 (b) 8 (c) 10 (d) 12**

7. If made up with threadless couplings, a 1" rigid metal conduit shall be supported at least every ____ feet.

**(a) 6 (b) 8 (c) 10 (d) 12**

## ANSWERS TO OPEN BOOK PRACTICE KEY TO USING NEC INDEX

1. **(b)** 200-10b ex. "Attachment plugs" is listed in the index but it does not lead you to the answer. The next step is to check the answers for a **key** word. "Identification" or "Polarized" from the index will both lead you to article 200-10b and the answer.

2. **(a)** 210-52a1. The key is to read the exact wording in the question. We space outlets 12' apart in a residence, but section 210-52a1 states that **no point along the floor line** in any wall space is more than 6' from an outlet.

3. **(a)** Table 310-13. The key is THHN, the **N** is a nylon **covering**, not insulation. 0.067 square inches is from Table 8 which is for **bare** conductors. 0.319 ohms per m/ft from Table 8 is for **aluminum** not copper.

4. **(a)** 410-80a. The key is to check the NEC index for "Electric discharge lighting", check each listing and 410-Q will lead you to the answer in article 410-80a.

5. **(c)** 210-6d1(b). "Ballasts" and "Electric-discharge lamps" are listed in the index but are no help. "Tunnel" is not even listed in the index. The key, "lighting fixture voltages" from the index will lead you to article 210-6 and the answer.

6. **(c)** 384-10. "Noninsulated" is not listed in the index. "Busbars" is listed, but of no help. The key, think of where you would find a busbar located, in an enclosure, a **panelboard**. Article 384 will lead you to the answer in 384-10.

7. **(c)** 346-12. The key word is **"threadless"** couplings. Section 346-12b1 states a conduit shall be supported at least every 10'. 346-12b states if made up with **threaded** couplings, you can use Table 346-12b2 for supports.

# CLOSED BOOK EXAM #1

## 50 QUESTIONS
## TIME LIMIT - 1 HOUR

**TIME SPENT** ☐ **MINUTES**

**SCORE** ☐ **%**

## JOURNEYMAN CLOSED BOOK EXAM #1

1. ___ can be generated.

I. Electricity II. Electrical energy

**(a) I only (b) II only (c) both I & II (d) neither I nor II**

2. The phenomenon whereby a circuit stores electrical energy is called ___.

**(a) inductance (b) capacitance (c) resistance (d) susceptance**

3. A general term including material, fittings, devices, appliances, fixtures, apparatus, and the like used as a part of, or in connection with, an electrical installation is ___.

**(a) premises wiring (system) (b) service equipment**
**(c) utilization equipment (d) equipment**

4. A switch intended for use in general distribution and branch circuits. It is rated in amperes, and it is capable of interrupting its rated current at its rated voltage, is a ____ switch.

**(a) bypass isolation (b) general use (c) isolating (d) transfer**

5. The permanent joining of metallic parts to form an electrically conductive path that will ensure electrical continuity and the capacity to conduct safely any current likely to be imposed is known as ___.

**(a) ordinary tap joint (b) scarf joint (c) britannia joint (d) bonding**

6. An instrument that is used to measure the diameter of a wire or cable to thousandths of an inch is a ___.

**(a) galvanometer (b) micrometer (c) hydrometer (d) ruler**

7. A ___ squirrel cage motor can be started at full voltage.

I. Design A II. Design B III. Design C IV. Design D

**(a) I only (b) I & II only (c) III & IV only (d) I, II, III or IV**

8. A ___ is a braking system for an electric motor.

I. friction braking  II. plugging  III. dynamic braking

**(a) I only  (b) III only  (c) I & III only  (d) I, II or III**

9. Rigid metal conduit is permitted for wiring in hazardous locations if the conduit is threaded and made up wrench tight with at least ___ full threads.

**(a) 4  (b) 5  (c) 7  (d) 9**

10. A circuit breaker that has purposely introduced into it a delay in the tripping action and which delay decreases as the magnitude of the current increases is a ___ circuit breaker.

**(a) inverse time  (b) adjustable  (c) control vented  (d) vented power**

11. It is the intent of this Code that factory installed internal wiring or the construction of equipment need not be inspected at the time of installation of the equipment, except to ___.

I. detect alterations
II. detect damage
III. detect insulation type

**(a) I only  (b) II only  (c) I & II only  (d) I, II & III**

12. A premises wiring system whose power is derived from a source such as a transformer that has no direct connection to the supply conductors originating in another system is a/an ___ system.

**(a) integrated  (b) separately derived  (c) interactive  (d) isolated**

13. Listed or labeled equipment shall be installed, used, or both, in accordance with any instructions included ___.

I. by the foreman
II. in the listing or labeling
III. with the equipment from the manufacturer

**(a) I only  (b) II only  (c) II & III only  (d) I, II and III**

14. Where conductors with an ampacity higher than the ampere rating or setting of the overcurrent device are used, the ___ shall determine the circuit rating.

**(a) conductor ampacity  (b) overcurrent device**
**(c) combined rating  (d) derated ampacity**

15. ___ are permitted to protect motor branch circuit conductors from overload.

I. Thermal relays II. Inverse time circuit breakers III. Time delay fuses

**(a) I only (b) II only (c) II &III only (d) I, II & III**

16. The power factor of an incandescent light bulb would be ___.

**(a) unity (b) 0.7 leading (c) 0.7 lagging (d) zero**

17. ___ is a pliable raceway.

I. EMT II. ENT III. PVC

**(a) I only (b) II only (c) I & III only (d) I, II, & III**

18. Flexible cords and cables shall be protected by ___ where passing through holes in covers, outlet boxes, or similar enclosures.

I. fittings II. bushings III. tie wraps

**(a) I only (b) II only (c) II & III only (d) I & II only**

19. A transformer would most likely have a ____ efficiency.

**(a) 60% (b) 70% (c) 80% (d) 90%**

20. When alternating current flows through a conductor, there is an inductive action that causes the current in the conductor to be forced toward the outer surface. The current is greater at the surface than at the center of the conductor, this ___ will cause the resistance in the conductor to increase due to the increased heating of the conductor.

**(a) capacitive effect (b) skin effect (c) conductive effect (d) outer effect**

21. A value assigned to a circuit or system for the purpose of conveniently designating its voltage class is ___.

**(a) nominal voltage (b) voltage to ground (c) voltage (of a circuit) (d) voltage$^2$**

22. A type of AC motor that runs at a constant speed and is used for such purposes as an electric clock motor is a ___ motor.

**(a) AC squirrel cage (b) AC induction (c) wound rotor induction (d) synchronous**

23. ___ is the resistance at the point of contact of two conductors or one conductor and another surface.

**(a) Conductor resistance** **(b) Contact resistance**
**(c) Resistance per M/ft** **(d) Resistance per K/ft**

24. ___ is/are classified as a conduit body.

I. LB fitting II. FS box III. LR fitting

**(a) I & II only (b) II only (c) II & III only (d) I & III only**

25. ___ raceways are **not** suitable to enclose conductors that are subject to physical damage.

**(a) Rigid metal conduit** **(b) Intermediate metal conduit**
**(c) PVC schedule 40** **(d) PVC schedule 80**

26. A low power factor in an industrial plant is most likely caused by ___.

**(a) insufficient resistive loads** **(b) insufficient inductive loads**
**(c) excessive resistive loads** **(d) lack of synchronous condenser**

27. Where lighting outlets are installed in interior stairways, there shall be a wall switch at each floor level to control the lighting where the difference between floor levels is ___ steps or more.

**(a) two (b) four (c) six (d) eight**

28. A voltage or current that is reversed at regular intervals is called ___ voltage or current.

I. direct II. steady state III. sinusoidal

**(a) I only (b) II only (c) III only (d) none of these**

29. Of the following ___ is a **false** statement.

**(a) The term kilowatt indicates the measure of power which is all available for work.**
**(b) The term kilovolt-amperes indicate the apparent power made up of an energy component and a wattless or induction component.**
**(c) In an industrial plant, low power factor is usually due to underloaded induction motors.**
**(d) The power factor of a motor is much greater at partial loads than at full load.**

30. It is generally not good practice to supply lights and motors from the same circuit because ___.

**(a) lamps for satisfactory service must operate within closer voltage limits than motors.**
**(b) overloads and short circuits are more common on motor circuits.**
**(c) when motors are started, the large starting current causes a voltage drop on the circuit and the lights will blink or burn dim**
**(d) all of these**

31. In general, motors are designed to operate in a maximum ambient temperature of ___ unless specifically designed for a higher temperature.

**(a) 60° C (b) 50° C (c) 45° C (d) 40° C**

32. A type of single phase motor that can be operated on either ac or dc is a ___ motor.

**(a) multispeed (b) capacitor-start (c) universal (d) repulsion-induction**

33. For screw shell devices with attached leads, the conductor attached to the screw shell shall be ___ in color.

**(a) white or gray (b) orange (c) green (d) black**

34. Branch circuit conductors shall have an ampacity not less than ___.

**(a) the load increased 125%**
**(b) 100% of the load to be served**
**(c) 80% of the load to be served**
**(d) 125% of the continuous load plus 80% of the noncontinuous load**

35. A switch intended for isolating an electric circuit from the source of power that has no interrupting rating, and it is intended to be operated only after the circuit has been opened by some other means is a/an ___.

**(a) isolating switch (b) bypass isolation switch (c) general use switch (d) transfer switch**

36. Raceways or cable trays containing electric conductors shall not contain ___.

I. pipe for steam II. tube for air III. pipe for water

**(a) I only (b) II only (c) III only (d) I, II or III**

37. Not readily accessible to persons unless special means for access are used is ___.

**(a) elevated (b) guarded (c) isolated (d) listed**

TH

38. After cutting a conduit, to remove the rough edges on both ends, the conduit ends should be ____.

**(a) sanded (b) shaped (c) burnished (d) ground**

39. The instrument used to indicate phase relation between current and voltage is the ____.

**(a) megger (b) power factor meter (c) voltmeter (d) galvanometer**

40. To calculate the va, one needs to know the ____.

**(a) voltage and current (b) impedance and conductance**
**(c) resistance and impedance (d) ohms and resistance**

41. You have an adjustable trip coil rated at 5 amps on a 200-amp switch.If you want the switch to trip at 120 amps, the trip coil should be set at ____.

**(a) 2 amps (b) 3 amps (c) 4 amps (d) 5 amps**

42. When an ammeter is disconnected from an in-service current transformer, the secondary terminals of the current transformer must be ____.

**(a) shorted (b) open (c) disconnected (d) grounded**

43. Reactance will cause the current in a circuit to vary only when ____.

**(a) AC current flows (b) DC current flows**
**(c) there is no resistance in the circuit (d) there is resistance in the circuit**

44. Motors of 1/3, 1/4, and 1/8 hp are connected in parallel. Those motors deliver a total of ____.

**(a) 1 hp (b) 7/8 hp (c) 17/24 hp (d) .07 hp**

45. Flexible cords and cables shall not be used ____.

**(a) for wiring of cranes and hoists (b) for prevention of the transmission of noise or vibration**
**(c) to run through holes in floors (d) simply to facilitate frequent interchange**

46. A fixture that weighs more than ____ shall be supported independently of the outlet box.

**(a) 25 pounds (b) 30 pounds (c) 50 pounds (d) 75 pounds**

47. The force which moves electrons from atom to atom through a closed conducting path is called ____.

**(a) flux (b) resistance (c) admittance (d) emf**

48. An advantage of a 240-volt system compared with a 120-volt system of the same wattage is ____.

**(a) reduced voltage drop (b) reduced power use**
**(c) large currents (d) lower electrical pressure**

49. A resistor has an indicated tolerance error of 10 percent. With a value of 1,000 ohms, the minimum resistance the resistor may have is ____.

**(a) 1,100 ohms (b) 990 ohms (c) 910 ohms (d) 900 ohms**

50. A transformer has a primary voltage of 120 volts and a secondary voltage of 480 volts. If there are 40 turns on the primary, the secondary contains ____.

**(a) 10 turns (b) 40 turns (c) 120 turns (d) 160 turns**

# CLOSED BOOK EXAM #2

## 50 QUESTIONS
## TIME LIMIT - 1 HOUR

**TIME SPENT** ☐ **MINUTES**

**SCORE** ☐ **%**

## JOURNEYMAN CLOSED BOOK EXAM #2 One Hour Time Limit

1. Frequency is measured in ____.

**(a) hertz** **(b) voltage** **(c) rpm** **(d) foot pounds**

2. Which of the following would cause the most power to be dissipated in the form of heat?

**(a) $X_L$** **(b) $X_C$** **(c) resonance** **(d) resistance**

3. ____ is the combined opposition to current by resistance and reactance.

**(a) Q** **(b) Z** **(c) $X_C$** **(d) $I^2R$**

4. An electrician in the industry would first check the ____ to correct a low power factor.

**(a) resistance** **(b) hysteresis** **(c) inductive load** **(d) reluctance**

5. Single conductor cable runs within a building are generally more common than multicable runs because _____.

**(a) of conduit fill** **(b) of the temperature**
**(c) the splicing is easier** **(d) the weight is evenly distributed**

6. ____ has the highest electrical breakdown strength and longest life over all other materials used for insulation.

**(a) Rubber insulation** **(b) Woven cloth**
**(c) Impregnated paper** **(d) Thermoplastic**

7. Voltage in a generator is produced by ____.

**(a) resonance** **(b) pressure** **(c) cutting lines of force** **(d) chemical**

8. To adjust the voltage generated by a constant speed DC generator, you would change the ____.

**(a) stator** **(b) slip rings** **(c) brushes** **(d) field current**

9. The generator which is best suited for electroplating power is a ____ generator.

**(a) split-phase** **(b) six pole** **(c) separately excited** **(d) compound**

10. To change the rotation of a DC motor you would ____.

**(a) reverse capacitor leads** **(b) reverse A1 and A2**
**(c) reverse commutator** **(d) reverse F1 and F2**

11. Frequency is determined by the ____ of an alternator.

I. size II. number of poles III. voltage IV. rotation speed of armature

**(a) II only (b) II and III only (c) II and IV only (d) I, II and IV only**

12. An example of a "made" electrode would be ___.

**(a) metallic water pipe** **(b) metal frame of a building**
**(c) concrete-encased** **(d) ground rod**

13. Illumination is measured in ____.

**(a) luminous flux (b) lumens (c) temperature (d) foot candles**

14. A motor enclosure designed and constructed to contain sparks or flashes that may ignite surrounding gas or vapor is called _____.

**(a) non-ventilated (b) encapsulated (c) explosion proof (d) water cooled**

15. The output of a 3ø transformer is measured in units called ____.

**(a) watts (b) volt-amps (c) impedance (d) turns-ratio**

16. Three horsepower is equivalent to ____ watts.

**(a) 764 (b) 2292 (c) 2238 (d) none of these**

17. Sometimes copper conductors are coated (tinned) to help prevent ___.

**(a) higher resistances (b) mechanical damage (c) capacitive reactance (d) chemical reaction**

18. A wheatstone bridge is used to measure ____ resistance.

I. low II. medium III. high

**(a) I only (b) II and III only (c) III only (d) II only**

19. To check voltage to ground, you would check from ____.

**(a) the breaker to the cabinet** **(b) hot to neutral**
**(c) the breaker to the grounding terminal** **(d) all of these**

20. The inductive action that causes current to flow on the outside surface of a conductor is known as the ____.

**(a) corona effect** **(b) skin effect** **(c) electrolitic action** **(d) DC reactance**

21. Electrical continuity is required by the electrical code for metallic conduit ____.

**(a) to assure equipment grounding** **(b) to reduce static electricity**
**(c) to reduce inductive heat** **(d) to trace electrical wiring**

22. The resistance of an open circuit is equal to ____.

**(a) less than one ohm (b) zero (c) infinity (d) none of these**

23. An electrical timer switch for lighting is normally connected in ___ with the lighting circuit being controlled.

**(a) series (b) parallel (c) sequence (d) tandem**

24. The definition of ampacity is ____.

**(a) the current-carrying capacity of conductors expressed in volt-amps**
**(b) the current-carrying capacity expressed in amperes**
**(c) the current-carrying capacity of conductors expressed in wattage**
**(d) the current in amperes a conductor can carry continuously under the conditions of use without exceeding its temperature rating**

25. The grounded conductor would connect to the ___ of a lampholder.

**(a) screw shell (b) filament (c) base contact (d) lead in wire**

26. A three-phase, 6-pole AC alternator 34 kva, on a Y-connected system. During one complete mechanical rotation (360°) will have ___ electrical rotations.

**(a) 1 (b) 1 1/2 (c) 3 (d) 12**

27. The voltage per turn of the primary of a transformer is ____ the voltage per turn of the secondary.

**(a) more than (b) the same as (c) less than (d) none of these**

28. A single concrete-encased electrode shall be augmented by one additional electrode if it does not have a resistance to ground of ___.

**(a) 25 ohms (b) 30 ohms (c) 50 ohms (d) not a Code requirement**

29. Which of the following is **not** true about alternating current?

**(a) develops eddy current**
**(b) it can be transformed**
**(c) is suitable for charging batteries**
**(d) interferes with communication lines**

30. On a 120v 1ø circuit, ground fault protection for personnel operates on the principal of unbalanced current between ___.

**(a) the grounded and ungrounded conductor**
**(b) the ungrounded conductors**
**(c) the grounding conductor and the neutral conductor**
**(d) the service disconnect and the branch circuit**

31. In a 3ø circuit, how many electrical degrees separate each phase?
**(a) 360 (b) 180 (c) 120 (d) 90**

32. ___ duty is a type of service where both the load and the time intervals may have wide variations.

**(a) Continuous**
**(b) Periodic**
**(c) Intermittent**
**(d) Varying**

33. The definition of ambient temperature is ___.

**(a) the temperature of the conductor**
**(b) the insulation rating of the conductor**
**(c) the temperature of the area surrounding the conductor**
**(d) the maximum heat the insulation can be used within**

34. As the power factor of a circuit is increased ____.

**(a) reactive power is decreased**
**(b) active power is decreased**
**(c) reactive power is increased**
**(d) both active and reactive power are increased**

35. Tinning rubber insulated twisted cable is done to ____.

**(a) make the strands stronger**
**(b) prevent chemical reactions between the copper and the rubber**
**(c) increase the resistance**
**(d) meet NEMA requirements**

36. A negatively charged body has ____.

**(a) excess of electrons (b) excess of neutrons (c) deficit of electrons (d) deficit of neutrons**

37. A fluorescent light that blinks "on" and "off" repeatly may in time ____.

**(a) cause the fuse to blow**
**(b) cause the switch to wear out**
**(c) cause the wire to melt**
**(d) result in damage to the ballast**

38. Electrical appliances are connected in parallel because it ____.

**(a) makes the operation of appliances independent of each other**
**(b) results in reduced power loss**
**(c) is a simple circuit**
**(d) draws less current**

39. What relationship determines the efficiency of electrical equipment?

**(a) The power input divided by the output**
**(b) The volt-amps x the wattage**
**(c) The va divided by the pf**
**(d) The power output divided by the input**

40. What is the formula to find watt hours?

**(a) E x T x 1000 (b) E x I x T (c) I x E x T/1000 (d) E x T x ø/1000**

41. Of the six ways of producing emf, which method is used the least?

**(a) pressure (b) solar (c) chemical action (d) friction**

42. The voltage produced by electromagnetic induction is controlled by ____.

**(a) the number of lines of flux cut per second**
**(b) eddy currents**
**(c) the size of the magnet**
**(d) the number of turns**

43. As the power factor of a circuit is increased ____.

**(a) reactive power is decreased**
**(b) active power is decreased**
**(c) reactive power is increased**
**(d) both active and reactive power are increased**

44. The breakdown voltage of an insulation depends upon ____ value of AC voltage.

**(a) r.m.s. (b) effective (c) peak (d) 1.732 of peak**

45. The AC system is preferred to the DC system because ____.

**(a) DC voltage cannot be used for domestic appliances**
**(b) DC motors do not have speed control**
**(c) AC voltages can be easily changed in magnitude**
**(d) high-voltage AC transmission is less efficient**

46. DC series motors are used in applications where ____ is required.

**(a) constant speed (b) high starting torque (c) low no-load speed (d) none of these**

47. Basically all electric motors operate on the principle of repulsion or ____.

**(a) magnetism (b) induction (c) resistance (d) capacitance**

48. A capacitor opposes ____.

**(a) both a change in voltage and current**
**(b) change in current**
**(c) change in voltage**
**(d) none of these**

49. The armature current drawn by any DC motor is proportional to the ____.

**(a) motor speed (b) voltage applied (c) flux required (d) torque applied**

50. The greatest voltage drop in a circuit will occur when the ____ the current flow through that part of the circuit.

**(a) greater (b) slower (c) faster (b) lower**

# CLOSED BOOK EXAM #3

## 50 QUESTIONS
## TIME LIMIT - 1 HOUR

**TIME SPENT** ☐ **MINUTES**

**SCORE** ☐ **%**

**JOURNEYMAN CLOSED BOOK EXAM #3** **One Hour Time Limit**

1. The electromotive force required to cause a current to flow may be obtained _____.

I. thermally II. mechanically III. chemically

**(a) I only (b) I and III only (c) II and III only (d) I, II and III**

2. Which of the following is **not** true?

**(a) A fluorescent fixture is more efficient than an incandescent fixture.**
**(b) Room temperature has an affect on the operation of a fluorescent lamp.**
**(c) Fluorescent fixtures have a good power factor with the current leading the voltage.**
**(d) The life of a fluorescent bulb is affected by starting and stopping.**

3. Resistance opposes the flow of current in a circuit and is measured in _____.

**(a) farads (b) joules (c) ohms (d) henrys**

4. Which of the following is true?

**(a) Wooden plugs may be used for mounting electrical equipment in concrete.**
**(b) The high-leg conductor of a 4-wire delta is identified blue in color.**
**(c) The minimum size service permitted by the Code for a residence is 100 amps.**
**(d) The ungrounded conductor is connected to the screw shell of a lampholder.**

5. Multiple start buttons in a motor control circuit are connected in _____.

**(a) series (b) parallel (c) series-parallel (d) none of these**

6. Which of the following is **not** true?

**(a) Feeder demand factors are applicable to household electric ranges.**
**(b) A green colored conductor can be used as an ungrounded circuit conductor.**
**(c) Insulated conductors #6 or smaller shall be white or gray, no marking tape permitted.**
**(d) All joints or splices must be electrically and mechanically secure before soldering.**

7. Special permission is _____.

**(a) granted by the electrical foreman on the job**
**(b) verbal permission by the inspector**
**(c) given only once on one blueprint change request**
**(d) the written consent of the authority having jurisdiction**

8. One million volts can also be expressed as ____.

**(a) 1 millivolt (b) 1 kilovolt (c) 1 megavolt (d) 1 microvolt**

9. Resistance in a circuit may be ____.

I. resistance of the conductors II. resistance due to imperfect contact

**(a) I only (b) II only (c) both I and II (d) neither I nor II**

10. Which of the following is **not** true?

**(a) All receptacles on 15 and 20 amp branch circuits must be of the grounding type.**
**(b) Splices and joints shall be covered with an insulation equivalent to the conductor insulation.**
**(c) The size of the conductor determines the rating of the circuit.**
**(d) All 15 and 20 amp receptacles installed in a dwelling bathroom shall have GFCI protection.**

11. A magnetic field is created around a conductor ____.

**(a) whenever current flows in the wire, provided the wire is made of magnetic material**
**(b) only when the wire carries a large current**
**(c) whenever current flows in the conductor**
**(d) only if the conductor is formed into a loop**

12. A universal motor has brushes that ride on the ____.

**(a) commutator (b) stator (c) inter-pole (d) field**

13. How many kw hours are consumed by 25 - 60 watt light bulbs burning 5 hours in a 120v circuit?

**(a) 1.5 (b) 180 (c) 7.5 (d) 75**

14. A dynamo is ____.

**(a) a pole line insulator**
**(b) a tool used to test dielectric strength**
**(c) a meter used for checking the R.P.M. of a motor**
**(d) a machine for converting mechanical energy into electrical energy**

15. Which of the following is/are generally used for field magnets?

I. copper II. steel III. wrought iron

**(a) I and II only (b) I and III only (c) II and III only (d) I, II and III**

16. The difference between a neutral and a grounded circuit conductor is ____.

**(a) only a neutral will have equal potential to the ungrounded conductor**
**(b) only a neutrals outer covering is white or natural gray**
**(c) only a neutral carries unbalanced current**
**(d) there is no difference**

17. The normal rotation of an induction motor is _____ facing the front of the motor. (The front of a motor is the end opposite the shaft).

**(a) clockwise (b) counterclockwise**

18. A function of a relay is to _____.

**(a) turn on another circuit**
**(b) produce thermal electricity**
**(c) limit the flow of electrons**
**(d) create a resistance in the field winding**

19. Which of the following is **not** true?

**(a) It is an electrical impossibility to have a circuit with only inductive reactance because the metallic wire has a resistance.**
**(b) The voltage of a circuit is the greatest effective difference of potential that exists between any two conductors of a circuit.**
**(c) The current is said to lag the voltage in a circuit that has only capacitive reactance.**
**(d) Power factor is the phase displacement of current and voltage in an AC circuit.**

20. Unity power factor, which means that the current is in phase with the voltage, would be _____.

**(a) .50 (b) .80 (c) 0.10 (d) 1.0**

21. Rheostats and potentiometers are types of _____ resistors.

**(a) film (b) variable (c) fixed (d) wirewound**

22. A laminated pole is _____.

**(a) one built up of layers or iron sheets, stamped from sheet metal and insulated**
**(b) used in transmission lines over 100kv**
**(c) a pole soaked in creosote**
**(d) found in the western part of the U.S.A.**

23. Which of the following is true?

**(a) Conductors of different systems may not occupy the same enclosure.**
**(b) Knife switches should be mounted in a horizontal position.**
**(c) 75 amps is a standard size fuse.**
**(d) Circuits are grounded to limit excess voltage to ground, which might occur from lightning or exposure to other higher voltage sources.**

24. Electrical power is a measure of ____.

**(a) work wasted (b) voltage (c) rate at which work is performed (d) total work performed**

25. What percentage of the maximum (peak) voltage is the effective (R.M.S.) voltage?

**(a) 100% (b) 70.7% (c) 63.7% (d) 57.7%**

26. A low factor is commonly caused by ____.

I. induction motors II. synchronous motors III. fluorescent lights

**(a) III only (b) II and III only (c) I and III only (d) I, II and III**

27. Which of the following is **not** true?

**(a) Conduit painted with enamel cannot be used outdoors.**
**(b) All AC phase wires, neutral and equipment grounding conductors if used, must be installed in the same raceway.**
**(c) PVC shall have a minimum burial depth of 24".**
**(d) EMT raceway can be installed in an air conditioning-space heating duct.**

28. Which of the following is **not** true?

**(a) Equal currents flow in the branches of parallel circuits.**
**(b) The total resistance of a parallel circuit is less than the smallest resistor in the circuit.**
**(c) The total current in a parallel circuit is the sum of the branch currents.**
**(d) In a parallel circuit, there is more than one path for the current flow.**

29. Hysteresis is ____.

**(a) the tool used to read the specific gravity of a battery**
**(b) the lagging of magnetism, in a magnetic metal, behind the magnetizing flux which produces it**
**(c) the opposite of impedance**
**(d) none of these**

30. The electric pressure of a circuit would be the ____.

**(a) voltage (b) amperage (c) resistance (d) wattage**

31. Permeability is ____.

**(a) the opposite of conductance**
**(b) a measure of the ease with which magnetism passes through any substance**
**(c) the total resistance to current flow**
**(d) the liquid substance in a battery**

32. The Wheatstone bridge method is used for accurate measurements of ____.

**(a) voltage (b) amperage (c) resistance (d) wattage**

33. When a circuit breaker is in the OPEN position ____.

I. you have a short in the ungrounded conductor
II. you have a short in the grounded conductror

**(a) I only (b) II only (c) either I or II (d) both I and II**

34. In solving series-parallel circuits, generally you would ____.

**(a) treat it as a series circuit**
**(b) reduce it to its simplest form**
**(c) assume that all loads are equal**
**(d) treat it as a parallel circuit**

35. A commutator is ____.

**(a) a ditching machine**
**(b) the inter-poles of a generator**
**(c) a device for causing the alternating currents generated in the armature to flow in the same direction in the external circuit**
**(d) a transformer with a common conductor**

36. Which of the following is true?

**(a) EMT may be threaded**
**(b) The "white" colored conductor connected to the silver colored post on a duplex receptacle on a 120v two-wire branch circuit is called the "neutral" conductor.**
**(c) Plastic water pipe is approved to be used for electrical conduit.**
**(d) The screw shell of a lampholder may support a fixture weighing 6 pounds.**

37. To fasten a box to a terra cotta wall you should use which of the following?

**(a) wooden plug (b) lag bolt (c) expansion bolt (d) toggle bolt**

38. If a 240 volt heater is used on 120 volts, the amount of heat produced will be ____.

**(a) twice as great (b) four times as great (c) 1/4 as much (d) the same**

39. Which of the following about a strap wrench is/are true?

I. you can turn pipe using one hand II. use in a tight corner III. use on different sizes of pipe

**(a) I only (b) II only (c) III only (d) I, II and III**

40. When soldering a joint, the flux is used to ____.

**(a) keep the wire cool (b) keep the surface clean**
**(c) lubricate the joint (d) maintain a tight connection**

41. The transferring of electrons from one material to another would be ____.

**(a) electrochemistry (b) static electricity (c) solar electricity (d) piezoelectricity**

42. A minimum thickness of ____ inch/inches of concrete over conduits and raceways should be used to prevent cracking.

**(a) 1 (b) 2 (c) 3 (d) 4**

43. Wire connectors are generally classified as ____ type(s).

I. thermal II. pressure

**(a) I only (b) II only (c) both I and II (d) neither I nor II**

44. One of the disadvantages of indenter or crimp connectors is ____.

**(a) they must be re-crimped at each annual maintenance inspection**
**(b) that special tools are required to make the joint**
**(c) eventually they will loosen**
**(d) they can only be used for copper conductors**

45. The usual service conditions under which a transformer should be able to carry its rated load are _____.

I. at rated secondary voltage or not in excess of 105% of the rated value
II. at rated frequency
III. temperature of the surrounding cooling air at no time exceeding 40°C (104°F) and average temperature of the surrounding cooling air during any 24-hour period not exceeding 30°C (86°F)

**(a) I only (b) II only (c) III only (d) I, II, and III**

46. Which of the following is **not** true?

**(a) An autotransformer may be used as part of the ballast for lighting circuits.**
**(b) A branch circuit can never be supplied through an autotransformer.**
**(c) The losses of the autotransformer are less than those of a two-coil transformer.**
**(d) Autotransformers may be used as starting compensators for AC motors.**

47. Conductors supplying two or more motors shall have an ampacity equal to the sum of the full-load current rating of all the motors plus ____ % of the highest rated motor in the group.

**(a) 25 (b) 80 (c) 100 (d) 125**

48. The symbol for a wye connection is ____.

**(a) Σ (b) Δ (c) ø (d) Y**

49. Which of the following meters is a wattmeter?

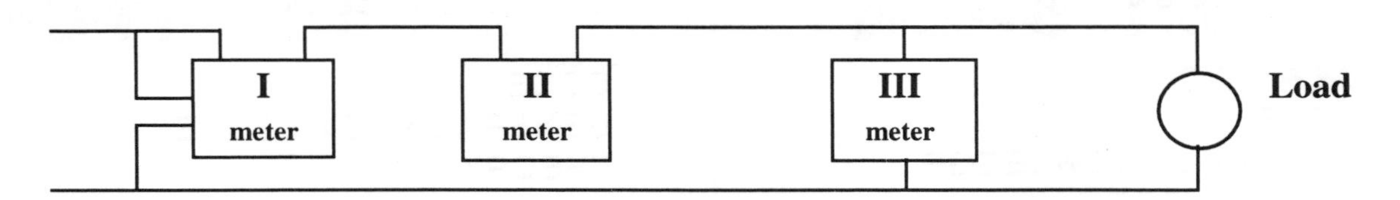


**(a) I only (b) II only (c) III only (d) I, II or III**

50. The voltage of a circuit is best defined as ____.

**(a) the potential between two conductors**
**(b) the greatest difference of potential between two conductors**
**(c) the effective difference of potential between two conductors**
**(d) the average RMS difference of potential between any two conductors**

# CLOSED BOOK EXAM #4

## 50 QUESTIONS
## TIME LIMIT - 1 HOUR

**TIME SPENT** ☐ **MINUTES**

**SCORE** ☐ **%**

## JOURNEYMAN CLOSED BOOK EXAM #4 One Hour Time Limit

1. Electrical current is measured in terms of ____.

**(a) electron pressure** **(b) electrons passing a point per second**
**(c) watts** **(d) resistance**

2. A stop switch is wired ____ in a motor circuit.

**(a) series (b) series-shunt (c) series-parallel (d) parallel**

3. An autotransformer has ____.

**(a) one coil (b) two coils (c) three coils (d) four coils**

4. What type of meter is shown below?

Shunt

**(a) wattmeter (b) ammeter (c) ohmmeter (d) voltmeter**

5. Concrete, brick or tile walls are considered as being ____.

**(a) isolated (b) insulators (c) grounded (d) dry locations**

6. ■ is the symbol for a ____ panel.

**(a) power (b) wall-mounted (c) lighting (d) surface-mounted**

7. A corroded electrical connection ____.

**(a) decreases the voltage drop** **(b) decreases the resistance of the connection**
**(c) increases the resistance at the connection** **(d) increases the ampacity at the connection**

8. An AC ammeter or voltmeter is calibrated to read RMS values; this means the meter is reading the ____ value.

**(a) maximum (b) peak (c) average (d) effective**

9. The correct connection for the two 120 volt lights to the single-pole switch would be ____.

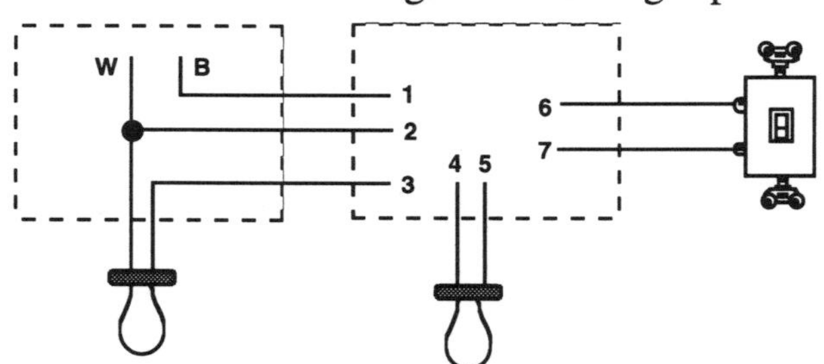


**(a) 1-4 2-6 3-5-7 (b) 1-6 2-5 3-4-7 (c) 1-7 2-5-6 3-4 (d) 1-5 2-6-7 3-4**

10. The location of a wall receptacle outlet in the bathroom of a dwelling shall be installed ____.

**(a) the Code does not specify the location** **(b) adjacent to the toilet**
**(c) within 36" of outside edge of basin** **(d) across from the shower**

11. On a delta three-phase four-wire secondary, how many hot wires may use the common neutral?

**(a) 1 (b) 2 (c) 3 (d) 4**

12. It shall be permissible to apply a demand factor of ____ to the nameplate-rating load of four or more appliances fastened in place served by the same feeder in a dwelling.

**(a) 70% (b) 75% (c) 60% (d) 80%**

13. Insulated nonmetallic boxes are made of ____.

I. polyvinyl chloride II. bakelite III. Bower-Barff lacquer

**(a) I only (b) II only (c) I and II only (d) I, II and III**

14. Tungsten-filament lamps can be used on ____ circuits.

**(a) AC (b) DC (c) AC and DC (d) none of these**

15. An overcurrent protective device with a circuit opening fusible part that is heated and severed by the passage of overcurrent through it is called a ____.

**(a) current-limiter (b) fuse (c) circuit breaker (d) thermal overload**

16. The service conductors between the terminals of the service equipment and a point usually outside the building, clear of building walls, where joined by tap or splice to the service drop is called the ____.

**(a) service drop (b) service-entrance conductors (c) service equipment (d) none of these**

17. If you needed to know the provisions for the installation of stationary storage batteries, you would refer to Article ____ of the Code.

**(a) 225 (b) 445 (c) 460 (d) 480**

18. A chain wrench can be used ____.

I. with one hand after the chain is around the conduit
II. in confined places and close to walls
III. for all sizes of conduit

**(a) I and II only (b) I and III only (c) II and III only (d) I, II and III**

19. To cut rigid conduit you should ___.

**(a) use 3-wheel pipe cutter**
**(b) use a cold chisel and ream the ends**
**(c) use hack saw and ream the ends**
**(d) order it cut to size**

20. A fixture that weighs more than ____ pounds shall be supported independently of the outlet box.

**(a) 25 (b) 30 (c) 35 (d) 50**

21. Is it permissible to install direct current and alternating current conductors in the same outlet box?

**(a) yes, if insulated for the maximum voltage of any conductor**
**(b) no, never**
**(c) yes, if the ampacity is the same for both conductors**
**(d) yes, in dry places**

22. Electrical equipment shall be installed ____.

**(a) better than the minimum Code allows**
**(b) according to the local Code when more stringent than the N.E.C.**
**(c) according to the N.E.C. regardless of local Code**
**(d) according to the local Code when less stringent than the N.E.C.**

23. Voltage drop in a wire is ___.

**(a) the wire resistance times the voltage**
**(b) a percentage of the applied voltage**
**(c) a function of insulation**
**(d) part of the load voltage**

24. Conductors shall **not** be installed in locations where the operating temperature will exceed that specified for the type of ____ used.

**(a) connectors (b) protection (c) insulation (d) wiring**

25. Galvanized conduit has a finish exterior and interior of ____.

**(a) lead (b) copper (c) nickel (d) zinc**

26. Which of the following is the best type of saw to use to cut a 3" diameter hole through 1/2" plywood?

**(a) circular saw (b) saber saw (c) hack saw (d) cross-cut saw**

27. Which of the following machine screws has the smallest diameter?

**(a) 6-32 x 1" (b) 10-32 x 3/4" (c) 8-32 x 1/2" (d) 10-24 x 3/8"**

28. Which of the following is the most important factor contributing to an electricians safety on the job?

**(a) work at a slow pace**
**(b) always wear leather gloves**
**(c) be alert at all times**
**(d) never be late for break**

29. A one-quarter bend in a raceway is equivalent to an angle of ____ degrees.

**(a) 90 (b) 45 (c) 25 (d) 180**

30. A 3Ω, a 6Ω, a 9Ω and a 12Ω resistor are connected in parallel. Which resistor will consume the most power?

**(a) 3Ω (b) 6Ω (c) 9Ω (d) 12Ω**

31. Listed ceiling (paddle) fans that do not exceed ____ pounds in weight, with or without accessories, shall be permitted to be supported by outlet boxes identified for such use.

**(a) 35 (b) 45 (c) 50 (d) 60**

32. The best way to lay out a 40 foot long straight line on a floor is to ____.

**(a) use a steel measuring tape with dark crayon**
**(b) use a plumb bob with long string**
**(c) use a long 2 x 4 and a lead pencil**
**(d) use a chalk line**

33. Silver is used on electrical contacts to ____.

**(a) avoid corrosion (b) improve efficiency (c) improve continuity (d) improve appearance**

34. Electricians should be familiar with the rules and regulations of their job mainly to ____.

**(a) eliminate overtime (b) increase wages (c) perform their duties properly (d) save time**

35. To determine if the raceway is truly vertical an electrician would use a ____.

**(a) plumb bob (b) transit level (c) square (d) level**

36. In order to prevent a safety hazard an electrician should never ____.

**(a) strike a hardened steel surface with a hardened steel hammer**
**(b) use a soft brass hammer to strike a soft brass surface**
**(c) strike a soft iron surface with a hardened steel hammer**
**(d) use a soft iron hammer to strike a hardened steel surface**

37. Service drop conductors not in excess of 600 volts shall have a minimum clearance of ____ feet over residential property and driveways, and those commercial areas not subject to truck traffic.

**(a) 10 (b) 12 (c) 15 (d) 18**

38. When conduit or tubing nipples having a maximum length not to exceed 24" are installed between boxes they shall be permitted to be filled ____ percent of its total cross-sectional area.

**(a) 31 (b) 40 (c) 53 (d) 60**

39. Before using rubber gloves when working on high voltage equipment the gloves should be ____.

**(a) cleaned inside and out** **(b) tested to withstand the high voltage**
**(c) oiled inside and out** **(d) brand new**

40. Stranded wire should be ____ before being placed under a screw head.

**(a) tinned (b) twisted together tightly (c) coated with an inhibitor (d) sanded**

41. A 3Ω, 6Ω, 9Ω and a 12Ω resistor are connected in series. The resistor that will consume the most power is the ____ ohm.

**(a) 3Ω (b) 6Ω (c) 9Ω (d) 12Ω**

42. What Article of the NEC refers to grounding?

**(a) 230 (b) 240 (c) 250 (d) 300**

43. The total of the following numbers 8 5/8", 6 1/4", 7 3/16" and 5 1/4" is ____.

**(a) 27 5/16" (b) 26 1/8" (c) 28 7/8" (d) none of these**

44. A fusestat is different than the ordinary plug fuse because a fusestat _____.

**(a) doesn't have threads** **(b) has left-hand threads**
**(c) has different size threads** **(d) has an aluminum screwshell**

45. The symbol ⌖ usually indicates a (an) _____.

**(a) switch (b) receptacle (c) ceiling outlet (d) exhaust fan**

46. A fuse on a 20 amp branch circuit has blown. The fuse is replaced with a 20 amp fuse and the fuse blows when the switch is turned on. The electrician should ______.

**(a) check the ground rod connection first** **(b) change to a circuit breaker**
**(c) install a 30 amp fuse** **(d) check the circuit for a problem**

47. To sharpen an electricians knife, you would use a _____ stone.

**(a) rubber (b) carborundum (c) rosin (d) bakelite**

48. The decimal equivalent of 3/16" is _____.

**(a) 0.125 (b) 0.1875 (c) 5.33 (d) none of these**

49. When drilling into a steel I-beam, the most likely cause for breaking a drill bit would be _____.

**(a) the drill bit is too dull** **(b) too slow a drill speed**
**(c) too much pressure on the bit** **(d) too much cutting oil on bit**

50. Which of the fuses is blown?

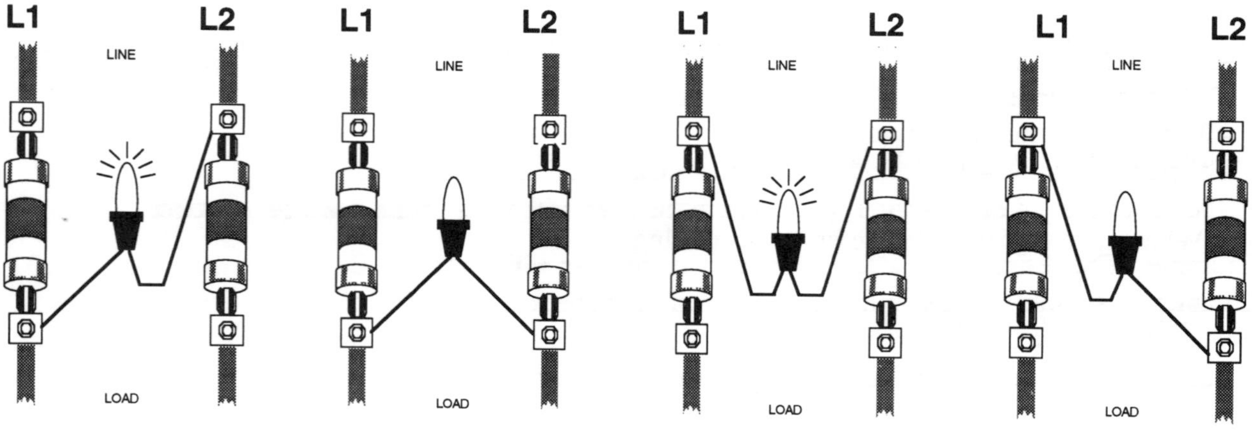

L1 L2 LINE LOAD | L1 L2 LINE LOAD | L1 L2 LINE LOAD | L1 L2 LINE LOAD

**(a) L1 fuse is blown (b) L2 fuse is blown (c) both fuses are blown (d) neither fuse is blown**

# CLOSED BOOK EXAM #5

## 50 QUESTIONS
## TIME LIMIT - 1 HOUR

**TIME SPENT** ______ **MINUTES**

**SCORE**  **%**

## JOURNEYMAN CLOSED BOOK EXAM #5 One Hour Time Limit

1. Locknuts are sometimes used in making electrical connections on studs. In these cases the purpose of the locknuts is to ____.

**(a) be able to connect several wires to one stud**
**(b) make it difficult to tamper with the connection**
**(c) make a tighter connection**
**(d) prevent the connection from loosening under vibration**

2. To cut rigid conduit you should ____.

**(a) use a 3-wheel pipe cutter**
**(b) use a cold chisel and ream the ends**
**(c) use a hacksaw and ream the ends**
**(d) order it cut to size**

3. In the course of normal operation the instrument which will be **least** effective in indicating that a generator may overheat because it is overloaded, is ____.

**(a) a wattmeter (b) a voltmeter (c) an ammeter (d) a stator thermocouple**

4. Two switches in one box under one face-plate is called a ____.

**(a) double-pole switch (b) two-gang switch (c) 2-way switch (d) mistake**

5. A conduit body is ____.

**(a) a cast fitting such as an FD or FS box**
**(b) a standard 10 foot length of conduit**
**(c) a sealtight enclosure**
**(d) a "LB" or "T", or similar fitting**

6. A dwelling unit is ____.

**(a) one unit of an apartment**
**(b) one or more rooms used by one or more persons**
**(c) one or more rooms with space for eating, living, and sleeping**
**(d) one or more rooms used as a housekeeping unit and having permanent cooking and sanitation provisions**

7. Enclosed means, surrounded by a ____ which will prevent persons from accidentally contacting energized parts.

I. wall II. fence III. housing or case

**(a) I only (b) II only (c) III only (d) I, II or III**

8. Where the conductor material is not specified in the Code, the conductors are assumed to be ____.

**(a) bus bars (b) aluminum (c) copper-clad aluminum (d) copper**

9. The voltage lost across a portion of a circuit is called the ____.

**(a) power loss (b) power factor (c) voltage drop (d) apparent va**

10. In a series circuit ____ is common.

**(a) resistance (b) current (c) voltage (d) wattage**

11. Batteries supply ____ current.

**(a) positive (b) negative (c) direct (d) alternating**

12. Electron flow produced by means of applying pressure to a material is called ____.

**(a) photo conduction (b) electrochemistry (c) piezoelectricity (d) thermoelectricity**

13. Raceways shall be provided with ____ to compensate for thermal expansion and contraction.

**(a) accordion joints (b) thermal fittings (c) expansion joints (d) contro-spansion**

14. An alternation is ____.

**(a) one-half cycle (b) one hertz (c) one alternator (d) two cycles**

15. What is the function of a neon glow tester?

I. Determines if circuit is alive
II. Determines polarity of DC circuits
III. Determines if circuit is AC or DC

**(a) I only (b) II only (c) III only (d) I, II and III**

16. What chapter in the Code is Mobile Homes referred to?

**(a) Chapter 3 (b) Chapter 5 (c) Chapter 6 (d) Chapter 8**

17. Never approach a victim of an electrical injury until you ____.

**(a) find a witness** **(b) are sure the power is turned off**
**(c) have a first-aid kit** **(d) contact the supervisor**

18. A wattmeter indicates ____.

I. real power II. apparent power if PF is not in unity III. power factor

**(a) I only (b) II only (c) III only (d) I, II and III**

19. The connection of a ground clamp to a grounding electrode shall be ____.

**(a) accessible (b) visible (c) readily accessible (d) in sight**

20. The current will lead the voltage when ____.

**(a) inductive reactance exceeds the capacitive reactance in the circuit**
**(b) reactance exceeds the resistance in the circuit**
**(c) resistance exceeds the reactance in the circuit**
**(d) capacitive reactance exceeds the inductive reactance in the circuit**

21. Mandatory rules of the Code are identified by the use of the word ____.

**(a) should (b) shall (c) must (d) could**

22. Which of the following is **not** one of the considerations that must be evaluated in judging equipment?

**(a) wire-bending and connection space** **(b) arcing effects**
**(c) longevity** **(d) electrical insulation**

23. To increase the range of an AC ammeter which one of the following is most commonly used?

**(a) a current transformer**
**(b) a condenser**
**(c) an inductance**
**(d) a straight shunt (not U-shaped)**

24. If a test lamp lights when placed in series with a condenser and a suitable source of DC, it is a good indication that the condenser is ____.

**(a) fully charged (b) short-circuited (c) open-circuited (d) fully discharged**

25. To transmit power economically over considerable distances, it is necessary that the voltages be high. High voltages are readily obtainable with ____ currents.

**(a) rectified (b) AC (c) DC (d) carrier**

26. Two 500 watt lamps connected in series across a 110 volt line draws 2 amperes. The total power consumed is ____ watts.

**(a) 50 (b) 150 (c) 220 (d) 1000**

27. The resistance of a copper wire to the flow of electricity _____.

**(a) decreases as the length of the wire increases**
**(b) decreases as the diameter of the wire decreases**
**(c) increases as the diameter of the wire increases**
**(d) increases as the length of the wire increases**

28. Enclosed knife switches that require the switch to be open before the housing door can be opened, are called ____ switches.

**(a) release (b) air-break (c) safety (d) service**

29. A type of cable protected by a spiral metal cover is called ____ in the field.

**(a) BX (b) greenfield (c) sealtight (d) Romex**

30. The resistance of a circuit may vary due to ____.

**(a) a loose connection (b) change in voltage (c) change in current (d) induction**

31. Grounding conductors running with circuit conductors may be ____.

I. uninsulated
II. a continuous green, if covered
III. continuous green with yellow stripe, if covered

**(a) I only (b) II only (c) III only (d) I, II and III**

32. For voltage and current to be in phase, ____.

I. the circuit impedance has only resistance
II. the voltage and current appear at their zero and peak values at the same time

**(a) I only (b) II only (c) both I and II (d) neither I nor II**

33. The definition of ampacity is ____.

**(a) the current-carrying capacity of conductors expressed in volt-amps**
**(b) the current-carrying capacity expressed in amperes**
**(c) the current-carrying capacity of conductors expressed in wattage**
**(d) the current in amperes a conductor can carry continuously under the conditions of use without exceeding its temperature rating**

34. Continuous duty is ____.

**(a) a load where the maximum current is expected to continue for three hours or more**
**(b) a load where the maximum current is expected to continue for one hour or more**
**(c) intermittent operation in which the load conditions are regularly recurrent**
**(d) operation at a substantially constant load for an indefinitely long time**

35. A location classified as dry may be temporarily subject to ____.

I. wetness II. dampness

**(a) I only (b) II only (c) both I and II (d) neither I nor II**

36. A ____ is an enclosure designed either for surface or flush mounting and provided with a frame, mat, or trim in which a swinging door or doors are or may be hung.

**(a) cabinet (b) panelboard (c) cutout box (d) switchboard**

37. A 15 ohm resistance carrying 20 amperes of current uses ____ watts of power.

**(a) 300 (b) 3000 (c) 6000 (d) none of these**

38. When using a #14-2 with ground Romex, the ground ____ carry current under normal operation.

**(a) will (b) will not (c) will sometimes (d) none of these**

39. As compared with solid wire, stranded wire of the same gauge size is ____.

**(a) better for higher voltages** **(b) given a higher ampacity**
**(c) easier to skin** **(d) larger in total diameter**

40. The type of AC system commonly used to supply both commercial light and power is the ____.

**(a) 3-phase, 3-wire (b) 3-phase, 4-wire (c) 2-phase, 3-wire (d) single-phase, 2-wire**

41. To make a good soldered connection between two stranded wires, it is **least** important to ____.

**(a) use enough heat to make the solder flow freely**
**(b) clean the wires carefully**
**(c) twist the wires together before soldering**
**(d) apply solder to each strand before twisting the two wires together**

42. The most important reason for using a condulet-type fitting in preference to making a bend in a one inch conduit is to ____.

**(a) avoid the possible flattening of the conduit when making the bend**
**(b) cut down the amount of conduit needed**
**(c) make a neater job**
**(d) make wire pulling easier**

43. When skinning a small wire, the insulation should be "penciled down" rather than cut square to ____.

**(a) allow more room for the splice**
**(b) save time in making the splice**
**(c) decrease the danger of nicking the wire**
**(d) prevent the braid from fraying**

44. Rubber insulation on an electrical conductor would quickly be damaged by continuous contact with ____.

**(a) water (b) acid (c) oil (d) alkali**

45. A tester using an ordinary light bulb is commonly used to test ____.

**(a) whether a circuit is AC or DC** **(b) for polarity of a DC circuit**
**(c) an overloaded circuit** **(d) for grounds on 120 volt circuits**

46. Pigtails are used on brushes to ____.

**(a) compensate for wear**
**(b) supply the proper brush tension**
**(c) make a good electrical connection**
**(d) hold the brush in the holder**

47. With respect to fluorescent lamps it is correct to state ____.

**(a) the filaments seldom burn out**
**(b) the starters and tubes must be replaced at the same time**
**(c) they are easier to install than incandescent light bulbs**
**(d) their efficiency is less than the efficiency of incandescent light bulbs**

48. A ____ stores energy in much the same manner as a spring stores mechanical energy.

**(a) resistor (b) coil (c) condenser (d) none of these**

49. An overcurrent trip unit of a circuit shall be connected in series with each ____.

**(a) transformer**
**(b) grounded conductor**
**(c) overcurrent device**
**(d) ungrounded conductor**

50. ____ lighting is a string of outdoor lights suspended between two points.

**(a) Pole (b) Festoon (c) Equipment (d) Outline**

# CLOSED BOOK EXAM #6

## 50 QUESTIONS
## TIME LIMIT - 1 HOUR

**TIME SPENT** ☐ **MINUTES**

**SCORE** ☐ **%**

## JOURNEYMAN CLOSED BOOK EXAM #6 One Hour Time Limit

1. Something that would effect the ampacity of a conductor would be ____.

I. voltage II. amperage III. length IV. temperature

**(a) I only (b) II only (c) III only (d) IV only**

2. Alternating currents may be increased or decreased by means of a ____.

**(a) motor (b) transformer (c) dynamo (d) megger**

3. Fixtures supported by the framing members of suspended ceiling systems shall be securely fastened to the ceiling framing member by mechanical means such as ____.

I. bolts or screws II. rivets III. clips identified for this use

**(a) I only (b) II only (c) III only (d) I, II or III**

4. Which has the highest electrical resistance?

**(a) brass (b) iron (c) water (d) paper**

5. Conductor sizes are expressed ____.

**(a) only in circular mils (b) in AWG or in circular mils**
**(c) in diameter or area (d) in AWG or millimeters**

6. Of the following, which one is **not** a type of file?

**(a) half round (b) bastard (c) tubular (d) mill**

7. Oil is used in many large transformers to ____.

**(a) prevent breakdown due to friction (b) lubricate the core**
**(c) cool and insulate the transformer (d) lubricate the coils**

8. Fractional horsepower universal motors have brushes usually made of ____.

**(a) copper strands (b) mica (c) carbon (d) thin wire rings**

9. When administering first aid to a worker suffering from fright as a result of falling from a ladder, the most important thing to do is ____.

**(a) position the person to a sitting position**
**(b) cover the person and keep the person warm**
**(c) apply artificial respiration immediately**
**(d) check the rungs of the ladder**

10. Which of the following would be used as a stop button?

**(a)** **(b)** **(c)** **(d)**

11. If a co-worker is burned by acid from a storage battery, the proper first aid treatment is to wash with ____.

**(a) iodine and leave it open to the air**
**(b) vinegar and apply a wet dressing**
**(c) water and apply vaseline**
**(d) lye and apply a dry bandage**

12. A type of motor that will **not** operate on DC is the ____.

**(a) series (b) short shunt (c) long shunt compound (d) squirrel cage**

13. Receptacles installed on ____ ampere branch circuits shall be of the grounding type.

**(a) 15 and 20 (b) 25 (c) 30 (d) 40**

14. Where conductors carrying alternating current are installed in metal enclosures or metal raceways, they shall be so arranged as to avoid heating the surrounding metal by induction, to accomplish this ____ shall be grouped together.

I. all phase conductors
II. where used, the neutral
III. all equipment grounding conductors

**(a) I only (b) I and II only (c) I and III only (d) I, II and III**

15. A(an) ____ changes AC to DC.

**(a) battery (b) capacitor (c) alternator (d) rectifier**

16. A steel measuring tape is undesirable for use around electrical equipment. The **least** important reason is the _____.

**(a) danger of entanglement in rotating machines** **(b) shock hazard**
**(c) short circuit hazard** **(d) magnetic effect**

17. _____ is the ability of a material to permit the flow of electrons.

**(a) Voltage (b) Current (c) Resistance (d) Conductance**

18. Automatic is self-acting, operating by its own mechanism when actuated by some impersonal influence, such as a change in _____.

I. temperature II. pressure III. current strength

**(a) I only (b) I and II only (c) II only (d) I, II and III**

19. A fitting is _____.

**(a) part of a wiring system that is intended primarily to perform an electrical function**
**(b) pulling cable into a confined area**
**(c) to be suitable or proper for**
**(d) part of a wiring system that is intended primarily to perform a mechanical function**

20. The neutral conductor _____.

**(a) is always the "white" grounded conductor**
**(b) has 70% applied for a household clothes dryer for a branch circuit**
**(c) never apply ampacity corrections**
**(d) carries the unbalanced current**

21. An appliance that is **not** easily moved from one place to another in normal use is a _____ appliance.

**(a) fastened in place (b) dwelling-unit (c) fixed (d) stationary**

22. All wiring must be installed so that when completed _____.

**(a) it meets the current-carrying requirements of the load**
**(b) it is free of shorts and unintentional grounds**
**(c) it is acceptable to Code compliance authorities**
**(d) it will withstand a hy-pot test**

23. Rosin is preferable to acid as a flux for soldering wire because rosin is ____.

**(a) a dry powder (b) a better conductor (c) a nonconductor (d) noncorrosive**

24. Utilization equipment is equipment which utilizes ____ energy for mechanical, chemical, heating, lighting or similar purposes.

I. chemical II. electric III. heat

**(a) I only (b) II only (c) III only (d) I, II and III**

25. The **main** purpose of using a cutting fluid when threading conduit is to ____.

**(a) prevent the formation of rust**
**(b) wash away the metal chips**
**(c) improve the finish of the thread**
**(d) prevent the formation of electrolytic pockets**

26. Of the following, the best indication of the condition of the charge of a lead acid battery is the ____.

**(a) temperature of the electrolyte**
**(b) level of the electrolyte**
**(c) open circuit cell voltage**
**(d) specific gravity**

27. In general, the most important point to watch in the operation of transformers is the ____.

**(a) core loss (b) exciting current (c) temperature (d) primary voltage**

28. When mounting electrical equipment, wooden plugs driven into holes in ____ shall **not** be used.

I. masonry II. concrete III. plaster

**(a) I only (b) II only (c) III only (d) I, II or III**

29. Mica is commonly used in electrical construction for ____.

**(a) commutator bar separators**
**(b) heater cord insulation**
**(c) strain insulators**
**(d) switchboard panels**

30. If a fuse becomes hot under normal load, a probable cause is ____.

**(a) excessive tension in the fuse clips**
**(b) rating of the fuse is too low**
**(c) insufficient pressure at the fuse clips**
**(d) rating of the fuse is too high**

31. For maximum safety the magnetic contactors used for reversing the direction of rotation of a motor should be _____.

**(a) operated from independent sources**
**(b) electrically interlocked**
**(c) mechanically interlocked**
**(d) electrically and mechanically interlocked**

32. Large squirrel cage induction motors are usually started at a voltage considerably lower than the line voltage to _____.

**(a) allow the rotor current to build up gradually**
**(b) permit starting under full load**
**(c) avoid excessive starting current**
**(d) obtain a low starting speed**

33. Which of the following is a motor starter?

**(a)** **(b)** **(c)** **(d)**

34. If the voltage on a light bulb is increased 10%, the bulb will _____.

**(a) fail by insulation breakdown**
**(b) have a longer life**
**(c) burn more brightly**
**(d) consume less power**

35. All edges that are invisible should be represented in a drawing by lines that are _____.

**(a) dotted (b) curved (c) solid (d) broken**

36. A light bulb usually contains _____.

**(a) air (b) neon (c) H2O (d) either a vacuum or gas**

37. The service disconnecting means shall be installed _____.

I. outside a building II. inside a building III. at the meter

**(a) I only (b) II only (c) III only (d) either I or II**

38. Critical burns are potentially _____.

**(a) life-threatening (b) disfiguring (c) disabling (d) all of these**

39. A set of lights switched from three different places can be controlled by ____ switch(es).

**(a) two 3-way and one 4-way** **(b) two 3-way and one 2-way**
**(c) 2 single-pole** **(d) four pole**

40. A fellow electrician is not breathing after receiving an electrical shock, but is no longer in contact with the electricity, the most important thing for you to do is _____.

**(a) start artificial respiration immediately** **(b) cover the person and keep warm**
**(c) move the person to a window** **(d) remove the persons shoes**

41. A wrench you would **not** use to connect rigid metal conduit is a ____ wrench.

**(a) box end** **(b) chain** **(c) strap** **(d) stillson**

42. The instrument that would prove **least** useful in testing for opens, grounds, and shorts after the wiring has been completed is the ____.

**(a) voltmeter** **(b) ammeter** **(c) ohmmeter** **(d) megger**

43. A stranded wire is given the same size designation as a solid wire if it has the same ____.

**(a) weight per foot** **(b) overall diameter**
**(c) strength** **(d) cross-sectional area**

44. A lighting fixture is to be controlled independently from two different locations. The type of switch required in each of the two locations is a ____.

**(a) double-pole, double-throw**
**(b) double-pole, single-throw**
**(c) single-pole, double throw**
**(d) single-pole, single-throw**

45.The rating "1000 ohms, 10 watts" would generally apply to a ____.

**(a) transformer** **(b) relay** **(c) resistor** **(d) heater**

46. The open circuit test on a transformer is a test for measuring its ____.

**(a) insulation resistance**
**(b) copper losses**
**(c) iron losses**
**(d) equivalent resistance of the transformer**

47. The proper way to open a knife switch carrying a heavy load is to ____.

**(a) open it with care, to avoid damage to the auxiliary blade by the arc**
**(b) open it slowly so that there will not be a flashover at the contacts**
**(c) tie a 5 foot rope on the switch handle and stand clear of the switch**
**(d) open it with a jerk so as to quickly break any arc**

48. When thermal overload relays are used for the protection of polyphase induction motors, their primary purpose is to protect the motors in case of ____.

**(a) short circuit between phases**
**(b) low line voltage**
**(c) reversal of phases in the supply**
**(d) sustained overload**

49. The National Electrical Code is sponsored by the ____.

**(a) Underwriters Lab**
**(b) National Safety Council**
**(c) National Electrical Manufacturers Association**
**(d) National Fire Protection Association**

50. Which of the following is an LB conduit body?

**(a)** 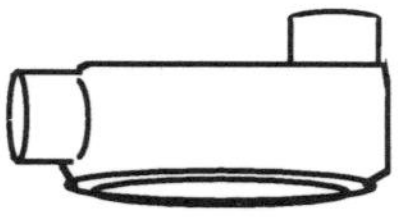 **(b)** 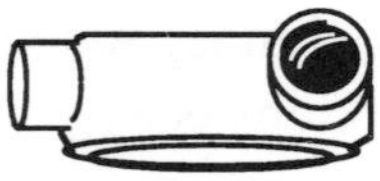 **(c)**  **(d)**

# CLOSED BOOK EXAM #7

## 50 QUESTIONS
## TIME LIMIT - 1 HOUR

**TIME SPENT**  **MINUTES**

**SCORE**  **%**

## JOURNEYMAN CLOSED BOOK EXAM #7

**One Hour Time Limit**

1. An advantage that rubber insulation has is that it _____.

**(a) is not damaged by oil**
**(b) is good for extreme temperatures**
**(c) does not absorb much moisture**
**(d) will not deteriorate with age**

2. The advantage of using a storage battery rather than a dry cell is the storage battery _____.

**(a) is portable (b) is less expensive (c) can be recharged (d) is easier to use**

3. The **least** desireable device for measuring an electrical cabinet containing live equipment is a _____.

**(a) 6' wooden ruler (b) plastic ruler (c) wood yardstick (d) 12' steel tape**

4. The relationship of a transformer primary winding to the secondary winding is expressed in _____.

**(a) wattage (b) turns-ratio (c) current (d) voltage**

5. When the size #12 of a stranded wire is referred to, this number specifies the:

**(a) strength of wire**
**(b) cross-sectional area of the wire**
**(c) square inch area of the insulation**
**(d) the pounds per square inch**

6. The purpose of a clip clamp is to _____.

I. ensure good contact between the fuse terminals of cartridge fuses and the fuse clips
II. make it possible to use cartridge fuses of a smaller size than that for which the fuse clips are intended
III. prevent the accidental removal of the fuse due to vibration

**(a) I, II and III (b) I only (c) II only (d) I and II only**

7. To increase the life of an incandescent light bulb you could _____.

**(a) use at a higher than rated voltage**
**(b) use at a lower than rated voltage**
**(c) turn off when not in use**
**(d) use at a higher wattage**

8. Which of the following statements about mounting single-throw knife switches in a vertical position is (are) correct?

I. The switch shall be mounted so that the blade hinge is at the bottom.
II. The supply side of the circuit shall be connected to the bottom of the switch.

**(a) I only (b) II only (c) both I and II (d) neither I nor II**

9. When re-routing conduit, it may be necessary to increase the wire size, if the distance is greater, in order to _____.

**(a) account for current drop**
**(b) allow for possible resistance drop**
**(c) compensate for voltage drop**
**(d) account for ampacity drop**

10. One megohm is the equivalent of _____.

**(a) 100 ohms (b) 1000 ohms (c) 100,000 ohms (d) 1,000,000 ohms**

11. On smaller guages of wire, they are pencil-stripped to prevent _____.

**(a) over stripping**
**(b) loosening of the wire-nut**
**(c) nicks in the wire**
**(d) other**

12. Galvanized conduit is made of _____.

**(a) iron (b) zinc (c) nickle (d) lead**

13. The frame of a motor is usually positively grounded to _____.

**(a) protect against shock**
**(b) remove the static currents**
**(c) provide 115 volts**
**(d) protect from lightning**

14. When wrapping a splice with both rubber and friction tape, the main purpose of the friction tape is to ______.

**(a) provide extra insulation (b) build up the insulation to the minimum thickness required**
**(c) protect the rubber tape (d) provide a waterproof seal**

15. An electrician should not wear shoes that have sponge rubber soles while working mainly because they ______.

**(a) wear out too quickly**
**(b) are not waterproof**
**(c) are not insulated**
**(d) are easily punctured when stepping on a nail**

16. The transformer output is measured by _____.

**(a) volts (b) amps (c) volt-amps (d) watts**

17. Which of the following hacksaw blades should be used for the best results in cutting EMT?

**(a) 12 teeth per inch (b) 18 teeth per inch**
**(c) 24 teeth per inch (d) 32 teeth per inch**

18. So constructed or protected that exposure to the weather will not interfere with successful operation is _____.

I. weather proof II. raintight III. watertight

**(a) I only (b) II only (c) I and II only (d) I, II and III**

19. The rating of the largest size regular plug fuse is _____ amperes.

**(a) 15 (b) 20 (c) 30 (d) 60**

20. A hacksaw with fine teeth used to cut raceways is commonly called a _____.

**(a) tube saw (b) keyhole saw (c) sabre saw (d) crosscut saw**

21. You shouldn't use a file without a handle because _____.

**(a) the file is hard to hold (b) the user may be injured**
**(c) the file will cut too deep (d) improper filing stroke**

22. The brightness of an incandescent lamp is rated in _____.

**(a) watts (b) foot candles (c) volt-amps (d) lumens**

23. If the primary winding of a 10 to 1 step down transformer has 20,000 turns, the secondary winding should have _____ turns.

**(a) 200,000 (b) 2000 (c) 200 (d) 20**

24. An electron is _____.

**(a) a neutron (b) an orbiting particle**
**(c) a proton (d) the smallest part of an atom with a negative charge**

25. The signals of electrical injury may include _____.

I. unconsciousness II. weak, irregular, or absent pulse III. dazed, confused behavior

**(a) I only (b) II only (c) III only (d) I, II or III**

26. This CODE is intended to be suitable for mandatory application by governmental bodies exercising legal jurisdiction over _____.

I. electrical installations II. and for use by insurance inspectors

**(a) both I and II (b) neither I nor II (c) I only (d) II only**

27. The name of the tool commonly used for bending small size conduit is a _____.

**(a) growler (b) mandrel (c) hickey (d) henry**

28. When cutting holes in masonry which of the following tools is most commonly used?

**(a) auger bit (b) router bit (c) star drill (d) reamer**

29. Electrician's diagonal lineman pliers should **not** be used to cut _____.

**(a) aluminum wire (b) copper wire (c) steel wire (d) copper-clad wire**

30. One of the following is the first thing to do when a person gets an electric shock and is still in contact with the supply:

**(a) remove the victim from contact by using a dry stick or dry rope**
**(b) treat for burns**
**(c) start artificial respiration immediately**
**(d) shut off power within 10 minutes**

31. A "mil" measures _____.

**(a) 1/8" (b) .000001" (c) .001" (d) .00010"**

32. The term "hertz" means _____.

**(a) car rental company (b) frequency (c) degrees (d) phase angle**

33. The difference of electrical potential between two conductors of a circuit is the:

**(a) resistance (b) amperage (c) voltage (d) wattage**

34. The letters DPDT are used to identify a type of _____.

**(a) insulation (b) fuse (c) motor (d) switch**

35. The term "ampere-hours" is associated with _____.

**(a) motors (b) transformers (c) electromagnets (d) storage batteries**

36. Which of the following would improve the resistance to earth?

I. Use multiple ground rods II. Treat the soil III. Lengthen the ground rod

**(a) I only (b) II and III only (c) I and III only (d) I, II and III**

37. A tap tool is a tool used to _____.

**(a) cut external threads (b) remove broken bolts**
**(c) ream raceways (d) cut internal threads**

38. When cutting a metal conduit with a hacksaw, the pressure applied to the hacksaw should be on _____.

**(a) the return stroke only (b) the forward stroke only**
**(c) both the forward and return stroke equally (d) none of these**

39. The switches to be closed in order to obtain a combined resistance of 5 ohms are _____ .

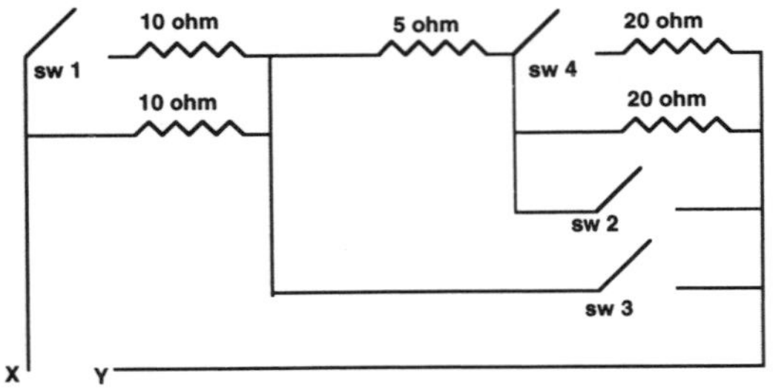

**(a) 1 and 3 (b) 2 and 3**

**(c) 1 and 2 (d) 1 and 4**

40. When the term "10-32" in connection with machine screws commonly used in lighting work, the number 32 refers to _____.

**(a) screw length (b) screw thickness (c) diameter of hole (d) threads per inch**

41. To fasten a box to a terra cotta wall you would use _____.

**(a) lag bolts (b) expansion bolts (c) wooden plugs (d) rawl plugs**

42. The output winding of a transformer is called the ____.

**(a) primary (b) output (c) secondary (d) both a & b**

43. The flux commonly used for the soldering of electrical conductors is ____.

**(a) zinc chloride (b) rosin (c) borax (d) none of these**

44. A shunt is sometimes used to increase the range of an electrical measuring instrument. The shunt is normally used when measuring ____.

**(a) AC voltage (b) DC voltage (c) DC amperes (d) AC amperes**

45. A battery operates on the principle of ____.

**(a) photo emission (b) triboelectric effect**
**(c) electro chemistry (d) voltaic conductivity**

46. When an electric current is forced through a wire that has considerable resistance, the ____.

I. ampacity will decrease II. voltage will drop III. wire will heat up

**(a) III only (b) I and II only (c) II and III only (d) I and III only**

47. The continuity of an electrical circuit can be determined in the field by the means of ____.

**(a) an ammeter (b) Wheatstone bridge (c) bell & battery set (d) wattmeter**

48. A wattmeter is connected in ____ in the circuit.

**(a) series (b) parallel (c) series-parallel (d) none of these**

49. A shunt is used to measure ____.

**(a) resistance (b) capacitance (c) current (d) wattage**

50. Which of the following is the symbol for a duplex outlet, split circuit?

R

**(a) (b) (c) (d)**

# CLOSED BOOK EXAM #8

## 50 QUESTIONS
## TIME LIMIT - 1 HOUR

**TIME SPENT** | | **MINUTES**

**SCORE** | | **%**

**JOURNEYMAN CLOSED BOOK EXAM #8** **One Hour Time Limit**

1. Using 1.5 volt dry cells, the voltage between A and B would be ____.

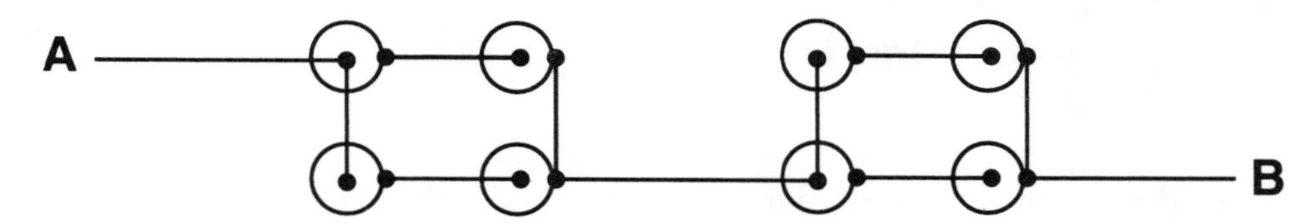


**(a) 1.5 (b) 4 (c) 6 (d) 12**

2. A rigid conduit connecting to an outlet box should have a ____.

**(a) bushing and locknut on the outside**
**(b) bushing on the outside and a locknut on the inside**
**(c) locknut and bushing on the inside**
**(d) locknut on the outside and a bushing on the inside**

3. Identified, as used in the Code in reference to a conductor or its terminals, means that such a conductor or terminal is to be recognized as ____.

**(a) grounded (b) bonded (c) colored (d) marked**

4. A toaster will produce less heat on low voltage because ____.

**(a) its total watt output decreases** **(b) the current will decrease**
**(c) the resistance has not changed** **(d) all of these**

5. If the current flow through a conductor is increased, the magnetic field around the conductor ____.

**(a) is unchanged (b) becomes stronger (c) collapses (d) becomes weaker**

6. Comparing a #6 conductor to a #10 conductor of equal lengths, the #6 will have lower ____.

**(a) cost (b) weight (c) resistance (d) strength**

7. The definition of ambient temperature is ____.

**(a) the temperature of the conductor**
**(b) the insulation rating of the conductor**
**(c) the temperature of the area surrounding the conductor**
**(d) the differential temperature**

8. The primary reason for using a hacksaw blade with fine teeth rather than coarse teeth when cutting large stranded conductors is ____.

**(a) a coarse blade would overheat the conductor**
**(b) a coarse blade breaks too easily**
**(c) to avoid snagging or pulling strands**
**(d) a fine blade will bend easier**

9. The standard residential service is a 3-wire, 240 volt single-phase system. The maximum voltage to ground in this system would be ____ volts.

**(a) 115 (b) 120 (c) 199 (d) 208**

10. When working on a motor, the electrician should ____ to prevent accidental starting of the motor.

**(a) remove the fuses**
**(b) ground the motor**
**(c) shut off the switch**
**(d) remove the belts**

11. It is the responsibility of the electrician to make sure his tools are in good condition because ____.

**(a) defective tools can cause accidents**
**(b) the boss may want to use them**
**(c) the company will pay for only one set of tools**
**(d) a good job requires perfect tools**

12. Continually overloading a conductor is a poor practice because it causes ____.

**(a) the conductor to melt**
**(b) the insulation to deteriorate**
**(c) the conductor to shrink**
**(d) damage to the raceway**

13. For better illumination you would ____.

**(a) random spacing of lights**
**(b) even spacing, numerous lights**
**(c) evenly spaced, higher ceilings**
**(d) cluster lights**

14. A junction box above a lay-in ceiling is considered ____.

**(a) concealed (b) accessible (c) readily accessible (d) recessed**

15. Which of the following metals is most commonly used in the filament of a bulb?

**(a) aluminum (b) mercury (c) tungsten (d) platinum**

16. Electrical equipment can be defined as ____.

I. fittings II. appliances III. devices IV. fixtures

**(a) I only (b) I and IV only (c) I, III and IV (d) all of these**

17. If two equal resistance conductors are connected in parallel, the resistance of the two conductors is equal to ____.

**(a) the resistance of one conductor**
**(b) twice the resistance of one conductor**
**(c) one-half the resistance of one conductor**
**(d) the resistance of both conductors**

18. Wire connection should encircle binding posts in the ____ manner the nut turns to tighten.

**(a) opposite (b) same (c) reverse (d) different**

19. Which of the following is a limit switch?

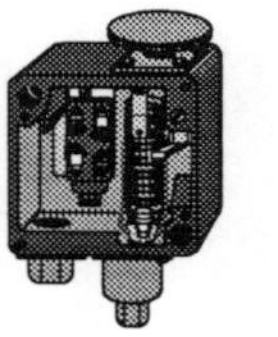

**(a)  (b)  (c)  (d)**

20. The primary and secondary windings of a transformer always have _____.

**(a) a common magnetic circuit**
**(b) the same size wire**
**(c) separate magnetic circuits**
**(d) the same number of turns**

21. Which of the following is **not** the force which moves electrons?

**(a) EMF (b) voltage (c) potential (d) current**

22. A motor with a wide speed range is a ____.

**(a) DC motor (b) AC motor (c) synchronous motor (d) induction motor**

23. The "stator" of an AC generator is another name for the ____.

**(a) rotating portion (b) slip rings (c) stationary portion (d) housing**

24. Where galvanized conduit is used, the main purpose of the galvanizing is to ____.

**(a) slow down rust (b) provide better continuity**
**(c) provide better strength (d) provide a better surface for painting**

25. To lubricate a motor sleeve bearing you would use ____.

**(a) grease (b) vaseline (c) oil (d) graphite**

26. When soldering conductors, flux is used ____.

**(a) to heat the conductors quicker**
**(b) to keep the surfaces clean**
**(c) to prevent loss of heat**
**(d) to bond the conductors**

27. ____ means so constructed or protected that exposure to the weather will not interfere with successful operation.

**(a) Weatherproof (b) Weather tight (c) Weather resistant (d) All weather**

28. The current used for charging storage batteries is ____.

**(a) square-wave (b) direct (c) alternating (d) variable**

29. You should close a knife switch firmly and rapidly as there will be less ____.

**(a) likelihood of arcing (b) wear on the contacts**
**(c) danger of shock (d) energy used**

30. If one complete cycle occurs in 1/30 of a second, the frequency is ____.

**(a) 30 hertz (b) 60 cycle (c) 115 cycle (d) 60 hertz**

31. An instrument that measures electrical energy is called the ____.

**(a) galvanometer (b) wattmeter (c) dynamometer (d) watthour meter**

32. In electrical wiring, "wire nuts" are used to ____.

**(a) connect wires to terminals** **(b) join wires and insulate the joint**
**(c) connect the electrode** **(d) tighten the panel studs**

33. Which of the following would be the best metal for a magnet?

**(a) steel (b) aluminum (c) lead (d) tin**

34. An electrician may use a megger ____.

**(a) to determine the RPM of a motor**
**(b) to determine the output of a motor**
**(c) to check wattage**
**(d) to test a lighting circuit for a ground**

35. The **least** important thing in soldering two conductors together is to ____.

**(a) use plenty of solder** **(b) use sufficient heat**
**(c) clean the conductors** **(d) use the proper flux**

36. The property of a circuit tending to prevent the flow of current and at the same time causing energy to be converted into heat is referred to as ____.

**(a) the inductance (b) the resistance (c) the capacitance (d) the reluctance**

37. Rigid conduit is generally secured to outlet boxes by ____.

**(a) beam clamps (b) locknuts and bushings (c) set screws (d) offsets**

38. Which one of the following is **not** a safe practice when lifting heavy items?

**(a) use the arm and leg muscles**
**(b) keep your back as upright as possible**
**(c) keep lifting a heavy object until you get help**
**(d) keep your feet spread apart**

39. A thermocouple will transform ____ into electricity.

**(a) current (b) heat (c) work (d) watts**

40. In a residence the wall switch controlling the ceiling light is usually _____.

**(a) connected across both lines**
**(b) a double pole switch**
**(c) connected in one line only**
**(d) a 4-way switch**

41. A switch which opens automatically when the current exceeds a predetermined limit would be called a _____.

**(a) limit switch (b) circuit breaker (c) DT disconnect (d) contactor**

42. A wattmeter is a combination of which two of the following meters?

I. ammeter II. ohmeter III. phase meter IV. volt meter V. power factor meter

**(a) II and III (b) I and V (c) I and IV (d) II and V**

43. What would the ohmmeter read ?

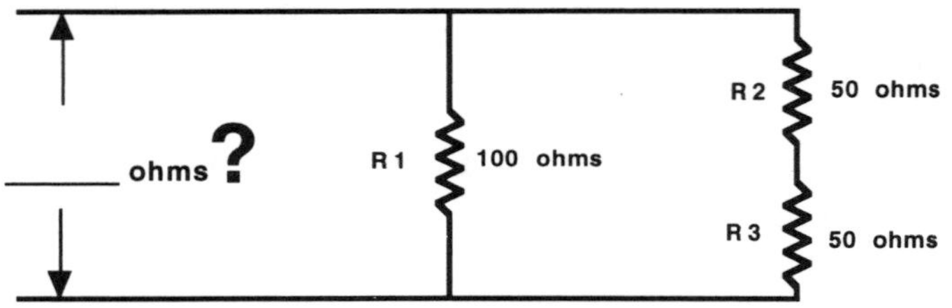

_____ ohms ?
R 1 100 ohms
R 2 50 ohms
R 3 50 ohms

**(a) 100 Ω (b) 200 Ω (c) 125Ω (d) 50 Ω**

44. Acid is not considered a good flux when soldering conductors because it _____.

**(a) smells bad (b) is corrosive (c) is non-conductive (d) costs too much**

45. If the spring tension on a cartridge fuse clip is weak, the result most likely would be _____.

**(a) the fuse would blow immediately**
**(b) the fuse clips would become warm**
**(c) the voltage to the load would increase**
**(d) the supply voltage would increase**

46. The branch-circuit loads specified by the Code for lighting and receptacles are considered _____.

**(a) minimum loads (b) maximum loads (c) loads to be served (d) peak loads**

47.The conductor with the highest insulation temperature rating is _____.

**(a) RH (b) TW (c) THWN (d) THHN**

48. After cutting a conduit, to remove the rough edges on both ends, the conduit ends should be _____.

**(a) reamed (b) filed (c) sanded (d) ground**

49. To fasten a raceway to a solid concrete ceiling, you would use _____.

**(a) toggle bolts (b) expansion bolts (c) wooden plugs (d) rawl plugs**

50. A commutator of a generator should be cleaned with which of the following?

**(a) emery cloth (b) graphite (c) a smooth file (d) fine sandpaper**

# CLOSED BOOK EXAM #9

## 50 QUESTIONS
## TIME LIMIT - 1 HOUR

**TIME SPENT** ☐ **MINUTES**

**SCORE** ☐ **%**

**JOURNEYMAN CLOSED BOOK EXAM #9** **One Hour Time Limit**

1. To control a ceiling light from five different locations it requires which of the following?

**(a) four 3-way switches and one 4-way switch**
**(b) three 4-way switches and two 3-way switches**
**(c) three 3-way switches and two 4-way switches**
**(d) four 4-way switches and one 3-way switch**

2. The advantage of AC over DC includes which of the following?

**(a) better speed control** **(b) lower resistance at higher current**
**(c) ease of voltage variation** **(d) impedance is greater**

3. Which of the following is considered the best electrical conductor?

**(a) iron wire (b) copper wire (c) aluminum wire (d) tin wire**

4. The liquid in a battery is called the _____.

**(a) askarel (b) festoon (c) hermetic (d) electrolyte**

5. A color code is used in multiple-conductor cables. For a 3-conductor cable the colors would be ____.

**(a) one black, one red and one white**
**(b) two black and one red**
**(c) one white, one black and one blue**
**(d) two red and one black**

6. Explanatory material in the Code is characterized by _____.

**(a) the word "shall" (b) FPN (c) the word "may" (d) the word "could"**

7. The identified grounded conductor of a lighting circuit is always connected to the screw of a light socket to _____.

**(a) reduce the possibility of accidental shock**
**(b) ground the light fixture**
**(c) improve the efficiency of the lamp**
**(d) provide the easiest place to connect the wire**

8. A ____ box may be weatherproof.

**(a) watertight (b) rainproof (c) raintight (d) all of these**

9. The Code requires that all AC phase conductors where used, the neutral and all equipment grounding conductors be grouped together when using metal enclosures or raceways. The principal reason for this is ____.

**(a) currents would circulate through individual raceways**
**(b) less expensive to install a single raceway**
**(c) less labor hours for pulling wires in a single raceway**
**(d) conductors are easier to pull in a single raceway**

10. Installing more than three current carrying conductors in the same conduit requires ____.

**(a) a larger conduit** **(b) high heat rated conductors**
**(c) derating of ampacity** **(d) continuous loading**

11. A ____ helps prevent arcing in movable contacts.

**(a) spring (b) condenser (c) resistor (d) hydrometer**

12. The ____ circuit is that portion of a wiring system prior to the final overcurrent protective device protecting the circuit.

**(a) service (b) feeder (c) power (d) branch**

13. When tightening a screw on a terminal, the end of the conductor should wrap around the screw in the same direction that you are turning the screw so that ____.

**(a) when you pull on the conductor it will tighten**
**(b) the screw will not become loose**
**(c) the conductor will act as a locking nut**
**(d) the conductor will not turn off**

14. Determining a positive wire on a single-phase circuit is ____.

**(a) possible with a wattmeter** **(b) possible with a voltmeter**
**(c) possible with an ammeter** **(d) an impossibility**

15. A ____ is used for testing specific gravity.

**(a) thermocouple (b) megger (c) hydrometer (d) galvanometer**

16. An autotransformer differs from other types of transformers in that ____.

**(a) its primary winding is always larger than its secondary winding**
**(b) it can be used only in automobiles**
**(c) its primary and secondary windings are common to each other**
**(d) it must be wound with heavier wire**

17. Where the ____ is likely to be high, asbestos insulation on the conductor would be a good choice.

**(a) temperature (b) humidity (c) voltage (d) amperage**

18. If the end of a cartridge fuse becomes warmer than normal, you should ____.

**(a) tighten the fuse clips**
**(b) lower the voltage on the circuit**
**(c) notify the utility company**
**(d) change the fuse**

19. Which of the following is the poorest conductor of electricity?

**(a) mercury (b) aluminum (c) carbon (d) silver**

20. The primary winding of a loaded step-down transformer has ____ compared to the secondary winding.

**(a) lower voltage and current** **(b) higher voltage and current**
**(c) higher voltage and lower current** **(d) lower voltage and higher current**

21. Copper is used for the tip of a soldering iron because ____.

**(a) copper will not melt** **(b) copper is a very good conductor of heat**
**(c) solder will not stick to other alloys** **(d) copper is less expensive**

22. The sum of the voltage drop around a circuit is equal to the source voltage is ____.

**(a) Kirchhoff's law (b) Ohm's law (c) Nevin's theory (d) Faraday's law**

23. Piezoelectric is caused by crystals or binding ____.

**(a) chemical (b) battery (c) pressure (d) heat**

24. Heavy-duty lampholders include ____.

**(a) admedium lampholders rated at 660 watts**
**(b) lampholders used on circuits larger than 20 amperes**
**(c) lampholders rated at not less than 750 watts**
**(d) all of the above**

25. The reason for installing electrical conductors in a conduit is ____.

**(a) to provide a ground**
**(b) to increase the ampacity of the conductors**
**(c) to protect the conductors from damage**
**(d) to avoid derating for continuous loading of conductors**

26. Discoloring of one end of a fuse normally indicates ____.

**(a) increased current (b) excessive voltage (c) low resistance (d) poor contact**

27. Wing nuts are useful on equipment where ____.

**(a) cotter pins are used** **(b) the nuts must be removed frequently**
**(c) a wrench cannot be used** **(d) screws cannot be used**

28. When resistors are connected in series, the total resistance is ____.

**(a) the sum of the individual resistance values**
**(b) the equivalent of the smallest resistance value**
**(c) the equivalent of the largest resistance value**
**(d) less than the value of the smallest resistance**

29. If a 120 volt incandescent light bulb is operating at a voltage of 125 volts, the result will be ____.

**(a) it may be enough to blow a fuse**
**(b) the bulb won't be as bright**
**(c) shorter life of the bulb**
**(d) the wattage will be less than rated**

30. Laminations are used in transformers to prevent ____.

**(a) copper loss (b) weight (c) eddy current loss (d) counter EMF**

31. The Code requires which of the following colors for the equipment grounding conductor?

**(a) white or gray** **(b) green or green with yellow stripes**
**(c) yellow** **(d) blue with a yellow stripe**

32. Sometimes mercury toggle switches are used in place of a regular toggle switch because they ____.

**(a) are easier to connect** **(b) do not wear out as quickly**
**(c) are less expensive** **(d) they glow in the dark**

33. The assigned color for the high-leg conductor of a three-phase, 4-wire delta secondary is ____.

**(a) red (b) black (c) blue (d) orange**

34. The Code rule for maximum 90 degree bends in a conduit between two boxes is four, the most likely reason for the total 360 degree limitation is ____.

**(a) it is unsafe**
**(b) it makes pulling the conductors through the conduit too difficult**
**(c) you can damage the galvanized coating on the conduit**
**(d) too many bends require extra wire to be pulled**

35. The correct word to define wiring which is not concealed is ____.

**(a) open (b) uncovered (c) exposed (d) bare**

36. A solenoid is a ____.

**(a) relay (b) permanent magnet (c) dynamo (d) electromagnet**

37. An electrician should always consider the circuit to be "hot" unless he definitely knows otherwise. The main reason is to avoid ____.

**(a) personal injury** **(b) having to find the panel**
**(c) saving time** **(d) shutting off the wrong circuit**

38. The best thing to cut PVC conduit within a tight area is ____.

**(a) a short hacksaw (b) a nylon string (c) a knife (d) a pipe cutter**

39. If a live conductor is contacted accidentally, the severity of the electrical shock is determined primarily by _____.

**(a) the size of the conductor** **(b) whether the current is DC or AC**
**(c) the current in the conductor** **(d) the contact resistance**

40. Ohm's law is _____.

**(a) an equation for determining power**
**(b) the relationship between voltage, current and power**
**(c) the relationship between voltage, current and resistance**
**(d) a measurement of wattage losses**

41. What is the normal taper on a standard conduit thread-cutting die?

**(a) 1/2" per foot (b) 1/4" per foot (c) 3/8" per foot (d) 3/4" per foot**

42. In an AC circuit the ratio of the power in watts to the total volt-amps is called the _____.

**(a) demand factor (b) power factor (c) turns-ratio (d) diversity factor**

43. The total load on any overcurrent device located in a panelboard shall not exceed _____ of its rating where in normal operation the load will continue for 3 hours or more.

**(a) 80% (b) 125% (c) 70% (d) 50%**

44. Four heaters, each having a resistance of 30 ohms, are connected in series across a 600-volt train circuit. The current is _____ amperes.

**(a) 5 (b) 17 (c) 20 (d) 80**

45. A ladder which is painted is a safety hazard mainly because the paint _____.

**(a) may conceal weak spots in the rails or rungs**
**(b) is slippery after drying**
**(c) causes the wood to crack more quickly**
**(d) peels and the sharp edges of the paint may cut the hands**

46. The chemical used as the agent in fire extinguishers to fight electrical fires is _____.

**(a) $CO_2$ (b) $K_OH$ (c) $H_2O$ (d) $L_O6$**

47. A location classified as ____ may be temporarily subject to dampness and wetness.

**(a) dry (b) damp (c) moist (d) wet**

48. The average dry cell battery gives an approximate voltage of ____.

**(a) 1.5 (b) 1.2 (c) 1.7 (d) 2.0**

49. The ____ circuit is that portion of a wiring system beyond the final overcurrent protection.

**(a) lighting (b) feeder (c) signal (d) branch**

50. What is the voltage between points **Y** and **Z** ?

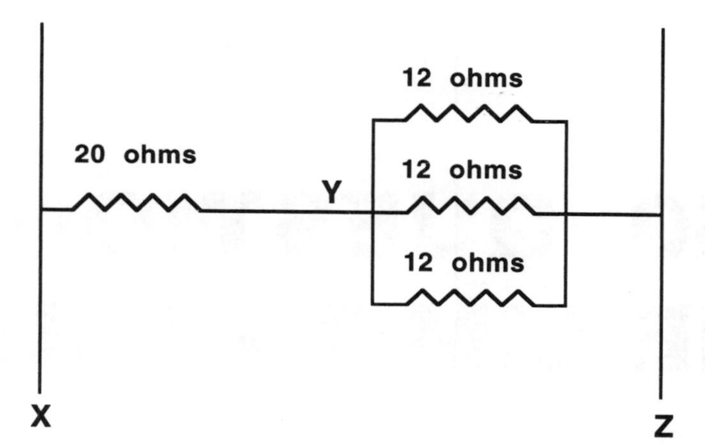


**EACH OF THE 12 OHM LOADS IS 2 AMPERES**

**(a) 72 volts (b) 120 volts (c) 24 volts (d) 144 volts**

# CLOSED BOOK EXAM #10

## 50 QUESTIONS
## TIME LIMIT - 1 HOUR

**TIME SPENT** ☐ **MINUTES**

**SCORE** ☐ **%**

**JOURNEYMAN CLOSED BOOK EXAM #10** **One Hour Time Limit**

1. The neutral conductor shall **not** be ____.

**(a) stranded (b) solid (c) insulated (d) fused**

2. The voltage drop in a line can be decreased by ____.

I. increasing the wire size
II. increasing the current
III. decreasing the load

**(a) I only (b) I and II only (c) I, II and III (d) I and III only**

3. In a residence, no point along the floor line in any wall space may be more than ____ feet from an outlet.

**(a) 6 (b) 6 1/2 (c) 12 (d) 10**

4. Insulating safety grips on tools ____.

**(a) are enough**
**(b) are not meant for that purpose**
**(c) should be used with other insulating equipment**
**(d) are not enough**

5. The rating of any one portable appliance shall not exceed ____ percent of the branch circuit rating.

**(a) 40 (b) 50 (c) 70 (d) 80**

6. A generic term for a group of non-flammable synthetic chlorinated hydrocarbons used as electrical insulating media:

**(a) askarel (b) acid (c) chloragorm (d) solder**

7. The part of an electrical system that performs a mechanical function rather than an electrical function is called a(n) ____.

**(a) receptacle (b) device (c) fitting (d) outlet**

8. An electrical condenser is best defined as ____.

**(a) a coil of wire**
**(b) a wrapping of layers of metal foil**
**(c) a coil of wire with layers of metal foil**
**(d) a wrapping of many layers of metal foil set apart by waxed paper**

9. Solid wire is preferred instead of stranded wire in panel wiring because ____.

**(a) costs less than stranded** **(b) solid will carry more current**
**(c) can be "shaped" better** **(d) no derating required for solid**

10. Which one of the following is not an insulator?

**(a) bakelite (b) oil (c) air (d) salt water**

11. The definition of accessible (wire):

**(a) admitting close approach**
**(b) not guarded by locked doors, elevation, etc.**
**(c) not permanently closed in by the building or structure**
**(d) all of the above**

12. The Code is designed for safety regardless of ____.

I. cost II. time III. maintenance IV. efficiency V. future expansion

**(a) I and II (b) III and IV (c) I through IV (d) I through V**

13. When voltage and current appear at their zero and peak values at the same time, they are in ____.

**(a) motion (b) group (c) phase (d) balanced**

14. What is meant by "traveler wires"?

**(a) wiring to a split receptacle** **(b) two-wires between 3-way switches**
**(c) wiring to a door bell** **(d) out of state electrician**

15. On a #4 drill bit, the #4 is determined by ____.

**(a) hardness (b) size (c) strength (d) length**

16. Wiring systems in wet locations should be ____.

**(a) placed so a permanent air space separates them from the supporting surface**
**(b) separated by insulated bushings**
**(c) separated by non-combustible tubing**
**(d) protected by a guard strip**

17. The best type of fire extinguisher for an electrical fire is a ____.

**(a) dry chemical extinguisher** **(b) soda-acid extinguisher**
**(c) foam extinguisher** **(d) carbon monoxide extinguisher**

18. "Thermally protected" appearing on the nameplate of a motor indicates that the motor is provided with a ____.

**(a) fuse (b) switch (c) breaker (d) heat sensing element**

19. A capacitor is a device that ____ energy.

**(a) produces (b) stores (c) opposes (d) increases**

20. When working near acid storage batteries, extreme care should be taken to guard against sparks, essentially to avoid ____.

**(a) overheating the electrolyte** **(b) an electric shock**
**(c) a short circuit** **(d) an explosion**

21. Which of the following statements is **incorrect**?

**(a) current flowing through a conductor causes heat**
**(b) the conduit of an electrical system should be grounded**
**(c) volt meters are connected in parallel in a circuit**
**(d) rectifiers change DC to AC**

22. When installing raceway systems, it is essential that they be ____.

**(a) rigidly supported as required** **(b) exposed**
**(c) concealed in walls** **(d) readily accessible**

23. Which of the following is a "handy" box?

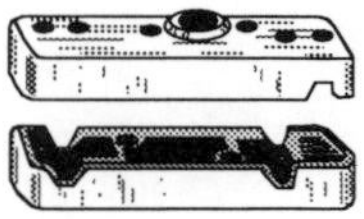

**(a)** 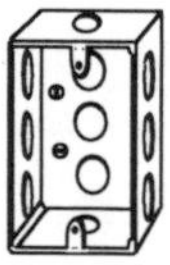 **(b)** 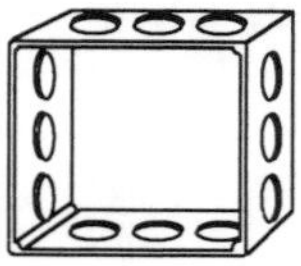 **(c)**  **(d)**

TH

24. The reason for grounding the frame of a portable electric hand tool is to ____.

**(a) prevent the frame of the tool from becoming alive to ground**
**(b) prevent overheating of the tool**
**(c) prevent shorts**
**(d) reduce the voltage drop**

25. Two metals of different materials shall not be joined together in order to avoid the ____ action.

**(a) rusting (b) galvanic (c) reverse (d) corrosion**

26. A ____ is a device which serves to govern in some predetermined manner the electric power delivered to the apparatus to which it's connected.

**(a) switch (b) feeder (c) service (d) controller**

27. The ungrounded conductor can be identified by the color ____.

**(a) white or gray (b) green or bare (c) pink flamingo (d) none of these**

28. What is the maximum number of overcurrent devices allowed in a lighting and appliance panelboard?

**(a) 24 (b) 30 (c) 36 (d) 42**

29. A ____ is a certain type cartridge fuse that can be readily replaced.

**(a) time-lag fuse** **(b) permanent fuse**
**(c) one-time fuse** **(d) renewable fuse**

30. The purpose of a Western Union splice is ____.

**(a) for the use of the utility companies only**
**(b) for the purpose of strengthening a splice**
**(c) for use on the west coast only**
**(d) none of these**

31. Electricity may be produced by means of ____ forces.

**(a) mechanical (b) thermal (c) chemical (d) all of these**

32. Copper-clad aluminum conductors have an ampacity ____.

**(a) lower than copper but higher than aluminum** **(b) equal to copper**
**(c) rating of their own** **(d) equal to aluminum**

33. The heating element in a toaster has a ____.

**(a) low resistance** **(b) high resistance**
**(c) high conductivity** **(d) none of these**

34. The total resistance of four 10 ohm resistors in parallel is ____.

**(a) 10 ohms (b) 2.5 ohms (c) 5 ohms (d) 4 ohms**

35. To mark a point on the floor directly beneath a point on the ceiling, it is best to use a ____.

**(a) transit rod (b) plumb bob (c) square (d) 12' tape**

36. Openings around electrical penetrations through fire-resistant rated walls, partitions, floors or ceilings shall be ____.

**(a) bushed (b) sleeved (c) firestopped (d) isolated**

37. A generator exciter uses ____ current.

**(a) alternating (b) direct (c) neither alternating nor direct (d) either alternating or direct**

38. When installing an instrument meter on a panel, to obtain accurate mounting ____.

**(a) use the meter and drill through the holes** **(b) drill oversize holes**
**(c) use a template** **(d) drill from back of panel**

39. The advantage of cutting a metal rigid conduit with a hacksaw rather than a pipe cutter is ____.

**(a) you do not need a vice** **(b) less energy required in cutting**
**(c) less reaming is required** **(d) threading oil is not required**

40. You would use an approved ____ to protect conductors from abrasion where they enter a box.

**(a) locknut (b) bushing (c) all thread (d) hickey**

41. To reverse the rotation of a three-phase motor you would ____.

**(a) reverse all the leads** **(b) reverse two of the four leads**
**(c) turn it around** **(d) reverse any two of the three leads**

42. The output rating of a one horsepower motor is ____.

**(a) 1840 watts (b) 746 watts (c) 1500 watts (d) 1000 watts**

43. In other than residential calculations, an ordinary outlet shall be calculated at ____.

**(a) 200 va (b) 600 watts (c) 300 watts (d) 180 va**

44. Impedance is present in the following type of circuit:

**(a) resistance (b) DC only (c) AC only (d) both AC and DC**

45. On an insulated conductor the type letter "TW" indicates ____.

**(a) tie-wire** **(b) thermoplastic-moisture resistant**
**(c) thermoplastic-waterproof** **(d) thermal-with nylon**

46. A load is considered to be continuous if it is expected to continue for ____.

**(a) 1/2 hour (b) 1 hour (c) 2 hours (d) 3 hours**

47. The standard classification of branch circuits applies only to those circuits with ____ outlets.

**(a) two or more (b) more than two (c) more than three (d) three or more**

48. If the primary of a transformer is 480 volts and secondary is 240/120v, the wire on the ____ is larger.

**(a) tertiary (b) secondary (c) primary (d) windings**

49. The important function of a type S fuse is ____.

**(a) non-interchangeable (b) slow burner (c) motor protection (d) fast acting**

50. If the voltage is doubled the ampacity of a conductor ____.

**(a) increases (b) decreases (c) doubles (d) remains the same**

# CLOSED BOOK EXAM #11

## 50 QUESTIONS
## TIME LIMIT - 1 HOUR

**TIME SPENT** ☐ **MINUTES**

**SCORE**  **%**

**JOURNEYMAN CLOSED BOOK EXAM #11** **One Hour Time Limit**

1. A ____ is a protective device for limiting surge voltages by discharging or bypassing surge current, and it also prevents continued flow of follow current while remaining capable of repeating these functions.

**(a) surge arrester (b) automatic fuse (c) fuse (d) circuit breaker**

2. A ____ conductor is one having one or more layers of non-conducting materials that are not recognized as insulation.

**(a) bare (b) covered (c) insulated (d) wrapped**

3. In a D.C. circuit, the ratio of watts to voltamperes is ____.

**(a) unity (b) greater than one (c) less than one (d) cannot tell what it might be**

4. A current limiting overcurrent protective device is a device which will ____ the current flowing in the faulted circuit.

**(a) reduce (b) increase (c) maintain (d) none of these**

5. The horsepower rating of a motor _____.

**(a) is a measure of motor efficiency** **(b) is the input to the motor**
**(c) cannot be changed to watts** **(d) is the output of the motor**

6. A common fuse and circuit breaker works on the principal that ____.

**(a) voltage develops heat** **(b) voltage breaks down insulation**
**(c) current develops heat** **(d) current expands a wire**

7. The voltage will lead the current when the ____ in the circuit.

**(a) inductive reactance exceeds the capacitive reactance**
**(b) reactance exceeds the resistance in the circuit**
**(c) resistance exceeds reactance**
**(d) capacitive reactance exceeds the inductive reactance**

8. Which of the following is an Allen head bolt?

**(a)** **(b)** **(c)** **(d)**

9. _____ is self-acting, operating by its own mechanism when actuated by some impersonal influence, as for example, a change in current strength, pressure, temperature, or mechanical configuration.

**(a) Remote-control (b) Automatic (c) Semi-automatic (d) Controller**

10. A 1000 watt, 120 volt lamp uses electrical energy at the same rate as a 14.4 ohm resistor on _____.

**(a) 120 volts (b) 115 volts (c) 208 volts (d) 240 volts**

11. When using compressed air to clean electrical equipment the air pressure should not exceed 50 pounds. The main reason is higher pressures _____.

**(a) may loosen insulating tape**
**(b) may blow dust to surrounding equipment**
**(c) introduce a personal hazard to the user**
**(d) may rupture the air hose**

12. Which of the following is **not** used to fasten equipment to concrete?

**(a) expansion bolt (b) lead shield (c) rawl plug (d) steel bushing**

13. A single-pole switch to operate a light will have the wiring connected in the _____ conductor.

**(a) grounded (b) identified (c) ungrounded (d) neutral**

14. The decimal equivalent of 9/16 is _____.

**(a) 0.5625 (b) 0.675 (c) 0.875 (d) none of these**

15. The information most useful in preventing the recurrence of a similar type accident when making out an accident report would be _____.

**(a) the nature of the injury**
**(b) the cause of the accident**
**(c) the weather conditions at the time**
**(d) the age of the person involved**

16. What is the total wattage of this circuit?

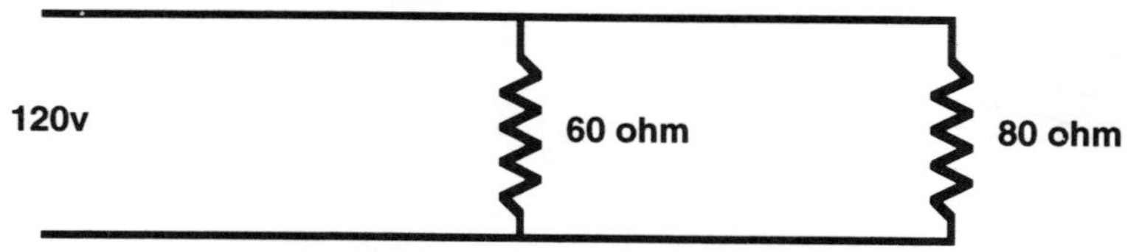

120v

60 ohm

80 ohm

**(a) 3.5 (b) 420 (c) 16,800 (d) 140**

17. Artificial respiration after a severe electrical shock is necessary when the shock results in _____.

**(a) broken limbs (b) bleeding (c) stoppage of breathing (d) unconsciousness**

18. If the circuit voltage is increased, all else remains the same, only the ____ will change.

**(a) resistance (b) current (c) ampacity (d) conductivity**

19. The two methods of making joints or connections for insulated cables are soldered connections and by means of solderless connection devices (wirenuts). The advantage(s) of a solderless connection (wirenut) is/are ____.

I. will not fail under short circuit due to melting of solder
II. mechanical strength as great as solder
III. reduces the time required to make a splice

**(a) I only (b) I and II only (c) II and III only (d) I, II and III**

20. Which of the following plugs is a polarized plug?

**(a) (b) (c) (d)**

21. When accidentally splashing a chemical into the eyes the best immediate first aid solution is to ____.

**(a) look directly into the sun** **(b) rub eyes with dry cloth**
**(c) flush eyes with clean water** **(d) close eyes quickly**

22. It is generally not good practice to supply lamps and motors from the same circuit because ____.

I. it is more economical to operate motors on a higher voltage than that of a lighting circuit
II. overloads and short circuits are more common on motor circuits and would put the lights out
III. when a motor is started it would cause the lights to dim or blink

**(a) I only (b) II only (c) III only (d) I, II and III**

23. Which of the following is the correct wiring to a light controlled by two 3-way switches?

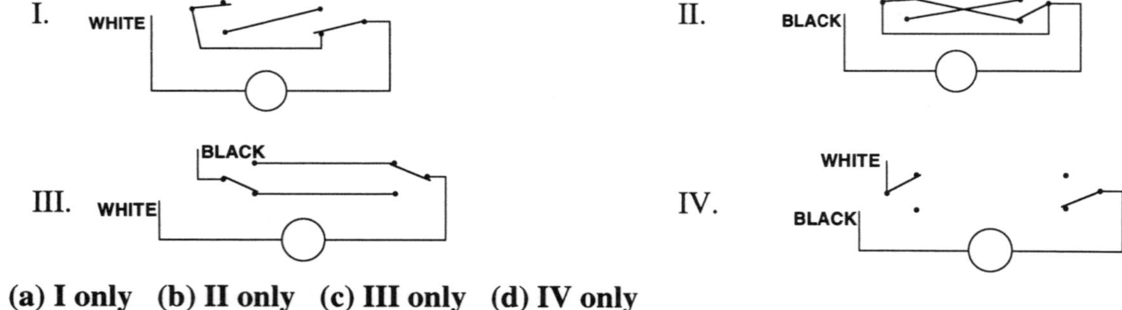


**(a) I only (b) II only (c) III only (d) IV only**

24. The Code considers low voltage to be _____.

**(a) 480 volts or less (b) 600 volts or less (c) 24 volts (d) 12 volts**

25. The cross-sectional area of the bus bar is ____ square inch.

1/4"
1/2"
10 3/4"

**(a) 0.125 (b) 1.34375 (c) 11.5 (d) none of these**

26. A high spot temperature in a corroded electrical connection is caused by a (an) ____.

**(a) increase in the flow of current through the connection**
**(b) decrease in the voltage drop across the connection**
**(c) increase in the voltage drop across the connection**
**(d) decrease in the effective resistance of the connection**

27. ____ is the symbol used for the delta connection.

**(a) Ω (b) Σ (c) ø (d) Δ**

28. Because aluminum is not a magnetic metal, there will be ____ present when aluminum conductors are grouped in a raceway.

**(a) no heat due to voltage**
**(b) no heating due to hysteresis**
**(c) no induced currents**
**(d) none of these**

29. A switch is a device for ____.

I. making or braking connections
II. changing connections
III. interruption of circuit under short-circuit conditions

**(a) I only (b) I and II only (c) II and III only (d) I, II and III**

30. At least two persons are required to be present during a high-voltage test because ____.

**(a) one person can cover while the one is on break**
**(b) high voltage is too heavy for one**
**(c) if one person is hurt the other person can help**
**(d) it eliminates overtime**

31. One of the essential functions of any switch is to maintain a ____.

**(a) good high-resistance contact in the closed position**
**(b) good low-resistance contact in the closed position**
**(c) good low-resistance contact in the open position**
**(d) good high-resistance contact in the open position**

32. Which of the following is a 30 amp receptacle?

**(a)** **(b)** **(c)** **(d)**

33. When the ground resistance exceeds the allowable value of 25 ohms, the resistance can be reduced by ____.

I. paralleling ground rods
II. using a longer ground rod
III. using a larger diameter ground rod
IV. chemical teatment of the soil

**(a) II and III only (b) I, II and III only (c) II, III and IV only (d) I, II, III and IV**

34. Silver and gold are better conductors of electricity than copper; however, the main reason copper is used is its ____.

**(a) weight (b) strength (c) melting point (d) cost is less**

35. Standard lengths of conduit are in 10 foot lengths. A required feeder raceway is 18 yards in length, how many lengths of 10 foot conduit would you need?

**(a) 4 (b) 5 (c) 6 (d) none of these**

36. The term "open circuit" means ____.

**(a) the wiring is in an open area**
**(b) the wiring is exposed on a building**
**(c) all parts of the circuit are not in contact**
**(d) the circuit has one end exposed**

37. Which of the items below is used to test specific gravity?

**(a)** **(b)** **(c)** **(d)**

38. Conduit should be installed as to prevent the collection of water in it between outlets. The conduit should not have a _____.

**(a) low point at an outlet** **(b) high point at an outlet**
**(c) high point between successive outlets** **(d) low point between successive outlets**

39. Brass is an alloy of ____.

**(a) zinc and copper (b) lead and copper (c) tin and lead (d) lead and tin**

40. Which type of the following portable fire extinguishers should be used on a live electrical fire?

**(a) carbon dioxide (b) water (c) foam (d) soda-acid**

41. Enclosed knife switches that require the switch to be open before the housing door can be opened, are called ____ switches.

**(a) release (b) air-break (c) safety (d) service**

42. Which of the following is a solenoid?

**(a) (b) (c) (d)**

43. What Article of the Code addresses high-voltage (over 600 volts)?

**(a) 450 (b) 230 (c) 680 (d) 490**

44. A close nipple ____.

**(a) is always 1/2" or less in length** **(b) has no threads**
**(c) has only internal threads** **(d) has threads over its entire length**

45. When applying rubber tape to an electrical splice, it would be necessary to ____.

**(a) stretch the tape properly during the application**
**(b) apply an adhesive to the splice before applying the tape**
**(c) apply the rubber tape after any other tape**
**(d) apply heat to the tape when installing**

46. A stranded wire with the same AWG as a solid wire ____.

**(a) is used for higher voltages**
**(b) has a higher ampacity**
**(c) is larger in total diameter**
**(d) has the same resistance**

47. A limit switch is used on a piece of machinery to open the circuit when the ____.

**(a) current exceeds a preset limit**
**(b) travel reaches a preset limit**
**(c) pressure exceeds a preset limit**
**(d) temperature reaches a preset limit**

48. With switches 1 and 2 closed the combined resistance of the circuit is ____ ohms.

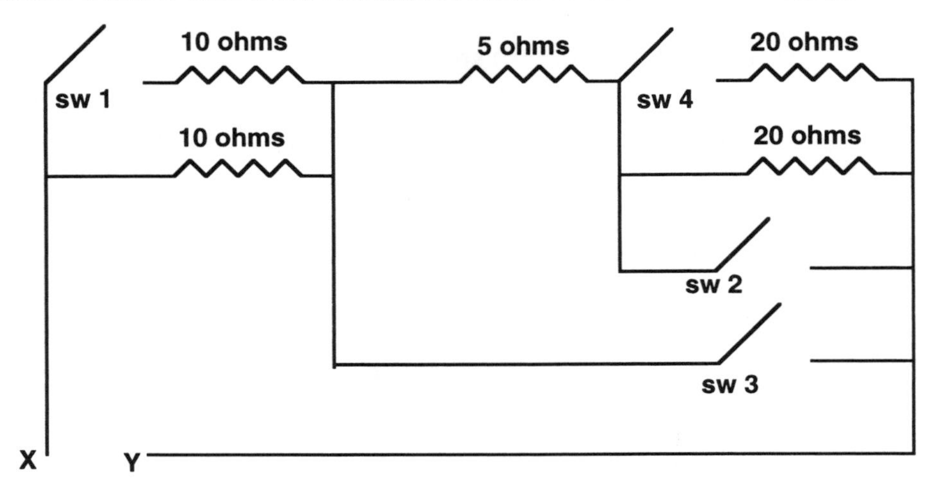


**(a) 30 (b) 25 (c) 10 (d) 3**

49. When rigid metal conduits are buried the minimum cover required by the Code is ____.

**(a) 6" (b) 12" (c) 18" (d) 24"**

50. A fixture that weighs more than ____ pounds shall not be supported by the screw shell of a lampholder.

**(a) 2 (b) 3 (c) 4 (d) 6**

# CLOSED BOOK EXAM #12

## 50 QUESTIONS
## TIME LIMIT - 1 HOUR

**TIME SPENT** ☐ **MINUTES**

**SCORE** ☐ **%**

## JOURNEYMAN CLOSED BOOK EXAM #12 — One Hour Time Limit

1. Your foreman asked you to measure the insulation resistance of some conductors. To do this you would use a ____.

**(a) hydrometer (b) megger (c) bell tester (d) wattmeter**

2. The main difference between a pipe thread and a machine thread is that the pipe thread is ____.

**(a) finer (b) longer (c) uneven (d) tapered**

3. Receptacles in residential wiring are regularly connected in ____.

**(a) parallel (b) perpendicular (c) series (d) diagonal**

4. A foreman in charge of a crew of men preparing to work on a low voltage tension circuit should caution them to ____.

**(a) work only when the load is zero**
**(b) consider the circuit hot at all times**
**(c) never work on any circuit alone**
**(d) wait until the circuit has been killed**

5. The term pneumatic refers to ____.

**(a) electricity (b) steam (c) air (d) oil**

6. What type of fastner would you use to mount a box to a hollow tile wall?

**(a) expansion bolts (b) toggle bolts**
**(c) rawl plugs (d) bolts with backing plates**

7. If a low resistance is connected in parallel with a higher resistance, the combined resistance is ____.

**(a) higher or lower than the low resistance depending on the size of the higher resistance**
**(b) always less than the low resistance**
**(c) always more than the higher resistance**
**(d) the total would be the low and high added together**

8. The lubricant used to make pulling wires through a conduit easier is ____.

**(a) grease (b) powdered pumice (c) vaseline (d) powdered soapstone**

9. The instrument by which electric power is measured is a ____.

**(a) ammeter (b) rectifier (c) voltmeter (d) wattmeter**

10. The connection between the grounded circuit conductor and the equipment grounding conductor at the service is called the ____ bonding jumper.

**(a) circuit (b) equipment (c) main (d) appliance**

11. The larger the conductor, the ____.

**(a) higher the resistance** **(b) lower the ampacity**
**(c) higher the voltage** **(d) lower the resistance**

12. A hook on the end of a fish tape is **not** to ____.

**(a) keep it from catching on joints and bends**
**(b) tie a swab to**
**(c) tie the wires, to be pulled**
**(d) protect the end of the wire**

13. Which of the following is a LL conduit body?

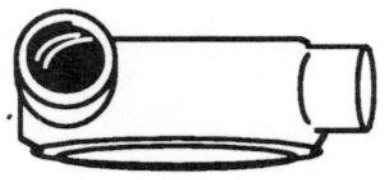

**(a)** 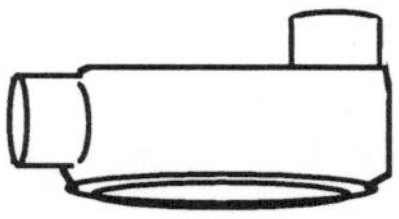 **(b)**  **(c)**  **(d)**

14. When soldering two copper conductors together, they are kept clean while heating by ____.

**(a) the use of flux**
**(b) applying the solder quickly**
**(c) rubbing often with emery cloth**
**(d) not permitting the open flame to touch them**

15. Metal cabinets used for lighting circuits are grounded to ____.

**(a) reduce shock hazard**
**(b) eliminate electrolysis**
**(c) assure that the fuse will blow in a defective circuit**
**(d) simplify the wiring**

16. In sockets, extension cord is protected by means of the ____ knot.

**(a) underwriters' (b) clove hitch (c) sheepshank (d) western union**

17. A branch circuit that supplies a number of outlets for lighting and appliances is a ____ branch circuit.

**(a) individual (b) multi-purpose (c) general purpose (d) utility**

18. When three equal resistors are connected in parallel, the total resistance is ____.

**(a) equal to the resistance of each (b) less than any one resistor**
**(c) greater than any one resistance (d) none of these**

19. The efficiency of a motor is a measure of ____.

**(a) the natural speed of the motor**
**(b) the torque the motor produces**
**(c) how well it converts electrical energy into mechanical energy**
**(d) the power output of the motor in horsepower**

20. When stripping insulation from an aluminum conductor ____.

I. remove insulation as you would sharpen a pencil
II. ring the conductor and slip the insulation off the conductor
III. peel the insulation back and then cut outwards

**(a) I, II and III (b) I and II only (c) I and III only (d) II and III only**

21. The ____ angle is the angle between the real power and the apparent power.

**(a) lag (b) power factor (c) voltage-current (d) watt**

22. The most heat is created when current flows through which of the following?

**(a) a 10 ohm condenser (b) a 10 ohm inductance coil**
**(c) a 10 ohm resistor (d) heat would be equal**

23. 60 cycle frequency travels 180 degrees in how many seconds?

**(a) 1/60 (b) 1/120 (c) 1/180 (d) 1/30**

24. The current-carrying capacity of conductors expressed in amperes is ____.

**(a) demand (b) pressure (c) ampacity (d) duty-cycle**

25. The electrician's tapered reamer is used for _____.

**(a) reaming the threads on couplings**
**(b) reaming the holes in bushings**
**(c) reaming the ends of rigid conduit after it is cut**
**(d) making holes in boxes**

26. Electricity is sold by the kilowatt which is _____ watts.

**(a) 10,000 (b) 1000 (c) 100 (d) 100,000**

27. Three-way switching does **not** use the following conductor:

**(a) ungrounded (b) traveler (c) grounded (d) switch leg**

28. The greater the number of free electrons the better the _____ of a metal.

**(a) insulation value (b) resistance (c) voltage drop (d) conductivity**

29. To cut Wiremold you would _____.

**(a) use a chisel**
**(b) use an approved cutter like an M.M. cutter**
**(c) use a pair of tin snips**
**(d) use a hacksaw and remove the burr with a file**

30. Electrical contacts are opened or closed when the electrical current energizes the coils of a device called a _____.

**(a) thermostat (b) reactor (c) condenser (d) relay**

31. A clamp-on ammeter will measure _____.

**(a) voltage when clamped on a single conductor**
**(b) current when clamped on a multi-conductor cable**
**(c) accurately only when parallel to cable**
**(d) accurately only when clamped perpendicular to a conductor**

32. When a current leaves its intended path and returns to the source bypassing the load the circuit is _____.

**(a) open (b) shorted (c) incomplete (d) broken**

33. The electric pressure or electromotive force is measured by the ____.

**(a) volt (b) electric meter (c) watt (d) kilowatt**

34. Conduit installed in a concrete slab is considered a ____.

**(a) damp location (b) moist location (c) wet location (d) dry location**

35. It is best as a safety measure, not to use water to extinguish electrical equipment fires. The main reason is that water ____.

**(a) may transmit shock to the user**
**(b) will turn to steam**
**(c) will not put the fire out**
**(d) may damage the wiring**

36. The total opposition to current flow in an AC circuit is expressed in ohms and is called ____.

**(a) impedance (b) conductance (c) reluctance (d) resistance**

37. Which of the items below is a rotometer?

**(a)  (b)  (c) 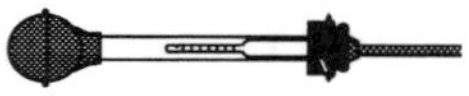 (d)**

38. When a person is burned the basic care steps are ____.

**(a) cover and cool the burned area** **(b) prevent infection**
**(c) care for shock** **(d) all of these**

39. A multimeter is a combination of ____.

**(a) ammeter, ohmmeter and wattmeter** **(b) voltmeter, ohmmeter and ammeter**
**(c) voltmeter, ammeter and megger** **(d) voltmeter, wattmeter and ammeter**

40. A good magnetic material is ____.

**(a) brass (b) copper (c) iron (d) aluminum**

41. Since fuses are rated by an amperage and voltage a fuse will work on ____.

**(a) AC only (b) AC or DC (c) DC only (d) any voltage**

42. A fuse puller is used in replacing ____.

**(a) cartridge fuses (b) plug fuses (c) link fuses (d) ribbon fuses**

43. A pendant fixture is a ____.

**(a) hanging fixture (b) recessed fixture (c) bracket fixture (d) none of these**

44. To fasten an outlet box between the studs in a wall constructed of metal lath and plaster, you would use ____.

**(a) cement or mortar (b) iron wire**
**(c) nylon lath twine (d) an approved box hanger**

45. The unit of measurement for electrical resistance to current is the ____.

**(a) watt (b) ohm (c) volt (d) amp**

46. A low energy power circuit ____.

**(a) is a remote-control circuit**
**(b) is a signal circuit**
**(c) has its power supplied by transformers and batteries**
**(d) none of these**

47. To convert AC or DC you will use a ____.

**(a) generator (b) rectifier (c) vibrator (d) auto-transformer**

48. $S_3$ is a symbol used on a drawing to indicate a ____ switch.

**(a) flush (b) single-pole (c) four-way (d) three-way**

49. Action requiring personal intervention for its control:

**(a) controller (b) automatic (c) periodic duty (d) non-automatic**

50. A voltmeter is connected in _____ with the load.

**(a) series (b) parallel (c) series-parallel (d) series-shunt**

# OPEN BOOK EXAM #1

## 50 QUESTIONS
## TIME LIMIT - 2 HOURS

**TIME SPENT** [ ] **MINUTES**

**SCORE** [ ] **%**

## JOURNEYMAN OPEN BOOK EXAM #1 Two Hour Time Limit

1. The minimum size service lateral to a branch circuit limited load is ___ copper.

**(a) #8 (b) #10 (c) #12 (d) none of these**

2. A household-type appliance with surface heating elements having a maximum demand of more than ___ amperes computed in accordance with Table 220-19 shall have its power supply subdivided into two or more circuits, each of which is provided with overcurrent protection rated at not over ___ amperes.

**(a) 40-40 (b) 50-40 (c) 50-60 (d) 60-50**

3. A 2400 volt lead cable can be bent up to ___ times its diameter.

**(a) 6 (b) 8 (c) 10 (d) 12**

4. A steel cable tray of .79 square inches is used as an equipment ground conductor. The maximum rating of the circuit breaker permitted for this application is ___ amps.

**(a) 1000 (b) 600 (c) 200 (d) 400**

5. Medium voltage cable insulation is rated for voltages ___ volts and higher.

**(a) 150 (b) 600 (c) 1000 (d) 2001**

6. A fixture rated at 7 amps requires a size ___ minimum fixture wire.

**(a) #16 (b) #18 (c) #14 (d) #12**

7. What is the minimum size THW copper-clad aluminum service entrance conductors for a calculated load of 182 amps to a 3-wire single phase dwelling unit?

**(a) #3/0 (b) #1/0 (c) #4/0 (d) #250 kcmil**

8. A bathroom in a dwelling has a counter space of seven feet including the sink. How many receptacles are required to serve this area?

**(a) 1 (b) 3 (c) 4 (d) none are required**

9. To ensure effective continuity between enclosures ___ shall be removed from the conduit threads.

**(a) ends (b) enamel (c) galvanize finish (d) aluminum**

10. An installation requires a device box with a capacity of 10.25 cubic inches. What is the minimum size box allowed?

**(a) 2" x 2" x 3"  (b) 3" x 2" x 2 1/4"  (c) 3" x 2" x 2"  (d) 2" x 3" x 3"**

11. The maximum percent of overcurrent protection allowed is ___ of the input current to an autotransformer when less than 9 amps.

**(a) 167%  (b) 150%  (c) 300%  (d) 125%**

12. A show window is calculated at ___ va per linear foot.

**(a) 180  (b) 1500  (c) 1800  (d) 200**

13. Aluminum fittings and enclosures shall be permitted to be used with ___.

**(a) both ferrous and nonferrous conduits**
**(b) PVC schedule 80 conduit**
**(c) electrical nonmetallic tubing**
**(d) steel electrical metallic tubing**

14. Type UF cable is manufactured in sizes #14 through # ___ copper.

**(a) 4/0  (b) 4  (c) 6  (d) 10**

15. Synchronous motors of the low torque, low speed type, such as are used to drive reciprocating compressors, pumps, etc., that start unloaded, do not require a fuse rating or circuit breaker setting in excess of ___ percent of full load current.

**(a) 150  (b) 200  (c) 250  (d) 400**

16. All 125 volt single phase receptacles within ___ feet of the inside walls of a hydromassage tub shall be protected by a ground fault circuit interrupter(s).

**(a) 5  (b) 10  (c) 12  (d) none of these**

17. Of the two to six service disconnecting means in a panel, only a disconnect used for ___ is permitted to be remote from the other disconnects.

**(a) control wiring**
**(b) a water pump intended for fire protection**
**(c) elevator panels**
**(d) supply to across the line starting**

18. A lighting fixture under a canopy is considered to be in a ___ location.

**(a) damp  (b) wet  (c) dry  (d) hazardous**

19. Resistors and reactors for use over 600 volts, shall not be installed in close enough proximity to combustible materials to constitute a fire hazard and shall have a clearance of not less than ___ from combustible materials.

**(a) 6" (b) 1' (c) 18" (d) 2'**

20. To reach a lighting fixture junction box you had to stand on a ladder. This junction box is considered to be ___.

**(a) concealed (b) readily accessible (c) accessible (d) hidden**

21. To settle a disagreement between an inspector and a contractor foreman, the ___ would have the final say.

**(a) local authority having jurisdiction**
**(b) local electrical board**
**(c) the IBEW**
**(d) the engineer**

22. The maximum number of 15 amp receptacles permitted on a free standing office partition is ___.

**(a) 10 (b) 13 (c) 2 (d) 6**

23. Transformer vaults shall have adequate structural strength and a minimum fire resistance of at least ___ hours. Unless protected by automatic sprinklers.

**(a) 6 (b) 1 1/2 (c) 3 (d) not required**

24. Flexible cords ___ and larger are used to supply approved appliances and are considered protected from overcurrent by overcurrent devices.

**(a) #18 (b) #16 (c) #14 (d) #12**

25. Panelboards, switches, gutters, wireways or transformers are permitted to be mounted above or below one another if ___.

**(a) rated 300v or less**
**(b) flush along the back edge**
**(c) they extend not more than 6 inches beyond the front of the equipment**
**(d) flush along the front edge**

26. In other than dwellings, ___ must have GFCI protection in a commercial building.

**(a) garage receptacle (b) outdoor receptacle (c) bathroom receptacle (d) none of these**

27. Size #18 or #16 fixture wires and flexible cords shall be permitted for the control and operating circuits of X-ray and auxiliary equipment where protected by not larger than ___ ampere overcurrent device.

**(a) 15 (b) 20 (c) 25 (d) 30**

28. Which of the following does not require a switched outlet according to the NEC?

**(a) walk through garage door (b) walk through porch door**
**(c) attic entrance (d) drive through garage door**

29. The highest current at rated voltage that a device is intended to interrupt under standard test conditions is know as ___.

**(a) overload (b) inverse time rated (c) thermal protector (d) interrupting rating**

30. Where fluorescent lighting fixtures are supported independently of the outlet box, they shall be connected by metal raceways, nonmetallic raceways or ___ may be used.

I. nonmetallic sheathed cable (romex) II. MI cable III. AC cable IV. MC cable

**(a) I and II only (b) II and III only (c) III only (d) I, II, III, and IV**

31. ___ is/are considered as service equipment by the NEC.

I. Meter socket enclosure II. Service disconnecting means III. Panelboard

**(a) I only (b) I and II only (c) II and III only (d) I, II, and III**

32. The sum of the diameters of all single conductors shall not exceed ___ when installed in a ventilated channel cable tray 4 inches inside width.

**(a) 2 inches (b) 3 inches (c) 4 inches (d) none of these**

33. Each autotransformer of 600 volts or less shall be protected by an individual overcurrent device installed in series with each ungrounded conductor and ___.

I. the overcurrent device shall be rated or set at not more than 125% of the rated full load input current of the autotransformer
II. an overcurrent device shall be installed in series with the shunt winding common to both the input and output circuits of the autotransformer

**(a) I only (b) II only (c) I or II (d) neither I nor II**

34. Where not listed for other support intervals, nonmetallic wireways shall be supported at maximum intervals of ___ feet.

**(a) 3 (b) 5 (c) 8 (d) 10**

35. A dry type transformer not rated over 112 1/2 kva installed indoors, shall have a separation of at least ___ inches from combustible material.

**(a) 24 (b) 18 (c) 12 (d) 6**

36. The residual voltage of a capacitor shall be reduced to ___ volts, nominal, or less with 1 minute after the capacitor is disconnected from the source of supply.

**(a) 0 (b) 15 (c) 30 (d) 50**

37. Where more than one building or other structure is on the same property and under single management, each building or other structure served shall be provided with means for disconnecting all ___ conductors located nearest the point of entrance of the supply conductors.

I. grounded II. ungrounded III. ungrounded and grounded

**(a) I only (b) II only (c) III only (d) I, II and III**

38. Where single phase loads are connected on the load side of a phase converter, they shall not be connected to the ___.

**(a) high leg (b) grounded phase (c) manufactured phase (d) neutral**

39. For an installation consisting of not more than two 2-wire branch circuits, the service disconnecting means shall have a rating of not less than ___ amperes.

**(a) 20 (b) 30 (c) 60 (d) 100**

40. The term pool includes swimming, wading and therapeutic pools and the term fountain includes ___.

I. ornamental pools II. drinking fountains III. display pools IV. reflection pools

**(a) I & II only (b) II & III only (c) III & IV only (d) I, III, & IV only**

41. Where the overcurrent device is rated over ___ amperes, the ampacity of the conductors it protects shall be equal to or greater than the rating of the overcurrent device.

**(a) 100 (b) 200 (c) 500 (d) 800**

42. When derating the ampacity of multiconductor cables to be installed in cable tray, the ampacity deration shall be based on ___.

I. the total number of current carrying conductors in the cable tray
II. the total number of current carrying conductors in the cable

**(a) I only (b) II only (c) either I or II (d) both I and II**

43. Where necessary to prevent ___, an automatic overcurrent device protecting service conductors supplying only a specific load, such as a water heater, shall be permitted to be locked or sealed where located so as to be accessible.

**(a) tripping (b) corrosion (c) heat build up (d) tampering**

44. An international term used to define a complete lighting unit consisting of a lamp or lamps together with the parts designed to distribute the light, to position and protect the lamps, and to connect the lamps to the power supply is a ___.

**(a) luminaire**
**(b) class I, division I light fixture**
**(c) class I, division II light fixture**
**(d) intrinsically safe light fixture**

45. A bonding jumper shall be used to connect the equipment grounding conductors of the derived system to the grounded conductor. This connection shall be made ___.

I. at any point on the separately derived system from the source to the first system disconnect
II. at any point on the separately derived system from the source to the first overcurrent device
III. at the source if the system has no disconnecting means or overcurrent device

**(a) I only (b) II only (c) III only (d) I, II or III**

46. A/an ___ shall be used to connect the grounding terminal of a grounding type receptacle to a grounded box.

**(a) neutral conductor**
**(b) branch circuit**
**(c) equipment bonding jumper**
**(d) bonding jumper main**

47. Thermoplastic-insulated fixture wire shall be durably marked with the AWG size, voltage rating and other required markings on the surface at intervals not exceeding ___ inches.

**(a) 6 (b) 12 (c) 18 (d) 24**

48. Fuses shall be plainly marked with ___.

I. ampere rating II. voltage rating III. interrupting rating where other than 10,000 amperes

**(a) I only (b) I & II only (c) I & III only (d) I, II & III**

49. Strut-type channel raceway shall be secured at intervals not exceeding ___ feet and within 3 feet of each outlet box.

**(a) 3 (b) 4 1/2 (c) 10 (d) 12**

50. Several motors, each not exceeding 1 horsepower in rating, shall be permitted on a nominal 120 volt branch circuit protected at not over __ amperes.

**(a) 15 (b) 20 (c) 30 (d) 40**

# OPEN BOOK EXAM #2

## 50 QUESTIONS
## TIME LIMIT - 2 HOURS

**TIME SPENT** ☐ **MINUTES**

**SCORE** ☐ **%**

**JOURNEYMAN OPEN BOOK EXAM #2 Two Hour Time Limit**

1. Service overhead conductors to a building or other structure (such as a pole) on which a meter of disconnecting means is installed shall be considered as a ___ and installed accordingly.

**(a) temporary service (b) service lateral (c) service drop (d) service point**

2. If a switch or circuit breaker serves as the disconnecting means for a permanently connected motor driven appliance of more than ___ horsepower, it shall be located within sight from the motor controller.

**(a) 1/8 (b) 1/4 (c) 1/2 (d) 3/4**

3. Overcurrent devices shall be enclosed in ___.

I. cabinets II. cutout boxes

**(a) I only (b) II only (c) I or II (d) none of these**

4. Where reduced heating of the conductors results from motors operating on duty-cycle, intermittently, or from all motors not operating at one time, the feeder conductors ____

**(a) are not allowed to have the ampacity reduced**
**(b) may have an ampacity less than specified if acceptable to the authority having jurisdiction**
**(c) must be sized no smaller than 125% of the largest motor connected to the feeder**
**(d) must be sized not smaller than 125% of the largest motor plus other loads**

5. Live parts of generators operated at more than ___ volts to ground shall not be exposed to accidental contact where accessible to unqualified persons.

**(a) 30 (b) 50 (c) 120 (d) 150**

6. A ___ is a circuit operating at 600 volts, nominal, or less, between phases that connects two power sources or power supply points, such as the secondaries of two transformers.

I. branch circuit individual II. branch circuit multiwire III. secondary tie

**(a) I only (b) II only (c) III only (d) I and II only**

7. Entrances to rooms and other guarded locations containing exposed live parts shall be marked with ___ warning signs forbidding unqualified persons to enter.

**(a) yellow (b) blue (c) conspicuous (d) orange**

8. Overhead spans of open conductors and open multiconductor cables not over 600 volts shall have a vertical clearance of not less than ___ feet above the roof surface.

**(a) 8 (b) 6 (c) 4 (d) 3**

9. Where single conductors #1/0 through 4/0 are installed in a ladder or ventilated trough cable tray they shall be installed in no more than ___.

I. a depth of 4" II. a depth of 6" III. a single layer

**(a) I only (b) II only (c) III only (d) I or II only**

10. Where flexible cords are permitted by the code to be permanently connected, it is permissible to omit ___ for such cords.

**(a) switches (b) receptacles (c) grounding connections (d) GFCI protection**

11. Listed or labeled equipment shall be installed, used, or both, in accordance with ___ .

**(a) the job specifications**
**(b) the plans**
**(c) the instructions given by the authority having jurisdiction**
**(d) the instructions included in the listing or labeling**

12. A grounding electrode connection that is encased in concrete or directly buried shall ___.

**(a) be made accessible**
**(b) be made only by exothermic welding**
**(c) be a minimum #4 bare**
**(d) not be required to be accessible**

13. Wet niche or no niche lighting fixtures that are supplied by a flexible cord or cable shall have all exposed noncurrent carrying metal parts grounded by an insulated copper equipment grounding conductor not smaller than the supply conductors and not smaller than ___.

**(a) #16 (b) #18 (c) #14 (d) #12**

14. If laid in notches in wood studs, joists, rafters, or other wood members ___ shall be protected against nails or screws by a steel plate at least 1/16" thick.

**(a) EMT (b) rigid nonmetallic conduit (c) intermediate steel conduit (d) flexible conduit**

15. A two pole circuit breaker that may be used for protecting a 3 phase corner grounded delta circuit shall be marked ___.

**(a) 1ø 120/240v (b) 1ø—3ø (c) 1ø/2ø/3ø (d) 480Y/277v**

16. When installing a surge arrester at the service of less than 1000 volts, the grounding conductor shall be connected to ___.

I. the grounded service conductor
II. the grounding electrode conductor
III. the grounding electrode for the service
IV. the equipment grounding terminal in the service equipment

**(a) I and II only (b) I and III only (c) III and IV only (d) I, II, III, or IV**

17. A means shall be provided in each metal box over 100 cubic inches for the connection of an equipment grounding conductor. The means shall be permitted to be ___.

I. a tapped hole II. the cover screw III. a screw used to mount the box

**(a) I only (b) II only (c) I and II only (d) I, II, or III**

18. A lighting fixture installed outdoors is permitted to be supported by ___.

I. trees II. a metal pole III. an outlet box

**(a) I only (b) II and III only (c) II only (d) I, II, or III**

19..The outer metal shell of a lampholder shall be lined with insulating material that shall prevent the shell and cap from becoming a part of the circuit. The lining shall not extend beyond the metal shell more than ___, but shall prevent any current carrying part of the lamp base from being exposed when a lamp is in the lampholding device.

**(a) 1/16" (b) 1/8" (c) 1/4" (d) 1/2"**

20. A ___ shall be used to connect the equipment grounding conductors, the service equipment enclosures, and where the system is grounded, the grounded service conductor to the grounding electrode.

**(a) bus bar** **(b) neutral conductor**
**(c) 5/8" ground rod** **(d) grounding electrode conductor**

21. For equipment rated 1200 amperes or more 600 volts or less, and over 6 feet wide, containing overcurrent devices, switching devices, or control devices, there shall be one entrance not less than ___ inches wide and 6 1/2 feet high at each end.

**(a) 24 (b) 30 (c) 36 (d) 48**

22. Appliances that have ___ that are to be connected by (1) permanent wiring method or (2) by field installed attachment plugs and cords with three or more wires (including the equipment grounding conductor) shall have means to identify the terminal for the grounded circuit conductor (if any).

I. screw shell lampholders II. single pole overcurrent device in the line III. single pole switch

**(a) I and II only (b) I and III only (c) II and III only (d) I, II and III**

23. Of the following, ___ box may be used for a floor receptacle.

**(a) a 4 11/16" x 1 1/4" square metal box with device ring listed for the purpose**
**(b) a 3" x 2" x 2 1/2" metal device box with device ring listed for the purpose**
**(c) a box listed specifically for this application**
**(d) any of these**

24. For a one family dwelling, at least one receptacle outlet, in addition to any provided for laundry equipment, shall be installed in each ___ .

I. basement II. detached garage with electric power III. attached garage

**(a) I only (b) II only (c) I and III only (d) I, II, and III**

25. Where nonmetallic sheathed cable is used with boxes no larger than ___ mounted in walls or ceilings and where the cable is fastened within 8 inches of the box, securing the cable to the box shall not be required.

**(a) 2 1/4" x 4" (b) 2/12" x 4" (c) 2" x 4" (d) 1 1/4" x 4"**

26. For swimming pool water heaters rated at more than ___ amperes that have specific instructions regarding bonding and grounding, only those parts designated to be bonded shall be bonded, and only those parts designated to be grounded shall be grounded.

**(a) 50 (b) 40 (c) 30 (d) 20**

27. Where a fixture is recessed in fire resistant material in a building of fire resistant construction, a temperature not higher than ___ shall be considered acceptable if the fixture is plainly marked that it is listed for that service.

**(a) 150°C (b) 165°C (c) 170°C (d) none of these**

28. A manufactured assembly designed to support and energize lighting fixtures that are capable of being readily repositioned is ___.

**(a) ceiling grid lighting (b) electric discharge lighting**
**(c) lighting track (d) open circuit lighting**

29. For AC adjustable voltage, variable torque drive motors, the ampacity of conductors, or ampere ratings of switches, circuit breakers or fuses and ground fault protection shall be based on the operating current marked on the nameplate. If the current does not appear on the nameplate, the ampacity determination shall be based on ___ of the values given in tables 430-149 and 430-150.

**(a) 80%   (b) 100%   (c) 125%   (d) 150%**

30. Which of the following is a **false** statement?

**(a) An accessible plug and receptacle shall be permitted to serve as the disconnecting means for a cord and plug connected appliance.**
**(b) For a household electric range, a plug and receptacle connection at the rear base is acceptable as the disconnect if it is accessible from the front by removal of a drawer.**
**(c) A counter mounted cooking unit shall be connected by a permanent wiring method.**
**(d) A switch with a marked off position that is a part of an appliance and disconnects all ungrounded conductors is permitted in a dwelling if the circuit is protected by a circuit breaker.**

31. Where a transformer vault is constructed with other stories below it, the floor shall have a minimum fire resistance of 3 hours unless ___.

**(a) the floors in contact with the earth not less than 3" thick**
**(b) protected with automatic sprinkler**
**(c) constructed of fire rated wallboard**
**(d) constructed of steel studs and fire rated wallboard**

32. A storage battery having the cells connected to operate at a voltage exceeding 250 volts but not over 600 volts, shall have insulation between groups and shall have a minimum separation between live battery parts of opposite polarity of ___ inch(es).

**(a) 2   (b) 1 1/2   (c) 1   (d) 1/2**

33. When calculating the conductor fill for strut-type channel raceway with internal joiners, the raceway shall be permitted to be filled to ___ percent of the cross-sectional area.

**(a) 20   (b) 25   (c) 30   (d) 40**

34. Which of the following wiring methods may be used inside the duct used for vapor removal and ventilation of commercial type cooking equipment?

**(a) nonmetallic sheathed cable   (b) EMT   (c) rigid steel conduit   (d) none of these**

35. Splices and taps shall be permitted in surface nonmetallic receways having a removable cover that is accessible after installation. The conductors, including splices and taps, shall not fill the raceway to more than ___ percent of its area at that point.

**(a) 31 (b) 40 (c) 53 (d) 75**

36. Cabinets and cutout boxes shall be deep enough to allow the closing of the doors when ___ ampere branch circuit panelboard switches are in any position; when combination cutout switches are in any position; or when other single throw switches are opened as far as their construction will permit.

**(a) 15 (b) 20 (c) 30 (d) 100**

37. Underfloor flat-top raceways over 4 inches but not over 8 inches wide with a minimum of 1 inch spacing between raceways shall be covered with concrete to a depth of not less than ___.

**(a) 3/4" (b) 1" (c) 1 1/2" (d) 2"**

38. Lighting fixtures located in the same room and not directly associated with a hydromassage bathtub, shall be installed in accordance with the requirements covering the installation of that equipment in ___.

**(a) swimming pool area (b) kitchen (c) exercise room (d) bathrooms**

39. The allowable fill for an 1 1/4 inch rigid schedule 40 PVC with more than 2 conductors is ___ sq. in.

**(a) .794 (b) .333 (c) .495 (d) .581**

40. Induction coils shall be prevented from inducing circulating currents in surrounding metallic equipment, supports, or structures by ___.

I. isolation II. shielding III. insulation of the current paths

**(a) I only (b) II only (c) III only (d) I, II or III**

41. At least one receptacle shall be located a minimum of 5 feet from and not more than ___ feet from the inside wall of a spa or hot tub installed indoors.

**(a) 6 (b) 10 (c) 12 (d) 20**

42. An electronically actuated fuse generally consists of all of the following EXCEPT ____?

**(a) a control module that provides current sensing**
**(b) electronically derived time-current characteristics**
**(c) an interrupting module that interrupts current when an overcurrent occurs**
**(d) a thermally sensitive part that is heated and severed by passage of overcurrent through it**

43. An underground pull box used for circuits of over 600 volts shall have the cover locked, bolted or screwed on, or the cover is required to weigh over ____ pounds.

**(a) 25 (b) 50 (c) 75 (d) 100**

44. Given: On a circuit where a grounding means does not exist, a nongrounding-type receptacle is replaced with a ground-fault circuit-interrupter-type (GFCI) receptacle which supplies no other receptacles. This new GFCI receptacle shall be marked ____.

**(a) "Not Grounded"**
**(b) "GFCI Protected"**
**(c) "No Equipment Ground"**
**(d) "No Grounded Conductor"**

45. Ground-fault circuit-interrupter (GFCI) protection is required in all of the following locations EXCEPT ____.

**(a) kitchen receptacles in an office building lunchroom installed within 6' of the sink**
**(b) kitchen receptacles in a dwelling installed to serve countertop surfaces 10' away from the sink**
**(c) receptacles in an office building restroom which has only a basin and toilet**
**(d) a receptacle provided for servicing a rooftop air conditioning unit on the roof of a warehouse**

46. For dwelling units, all of the following are true EXCEPT ____.

**(a) outdoor outlets are permitted to be supplied through the small appliance branch circuits**
**(b) the outlet for kitchen refrigeration equipment may be supplied by an individual 15 amp branch circuit**
**(c) bathroom receptacles shall be supplied by a 20 amp branch circuit which shall have no other outlets**
**(d) the clothes washer shall be supplied by a 20 amp branch circuit and outlets outside the laundry area are NOT permitted on this circuit**

47. In a recreational vehicle park, tent sites equipped with only 20 ampere supply facilities shall be calculated on the basis of ___ per site.

**(a) 180 va (b) 300 va (c) 360 va (d) 600 va**

48. Where GFCI protection is located in the power supply cord for an outdoor portable sign, the ground-fault circuit interrupter shall be located within ___ inches of the attachment plug.

**(a) 6**
**(b) 12**
**(c) 18**
**(d) 24**

49. Given: A fixed electric space heater without a motor is installed in a multifamily dwelling. The heater has no supplementary overcurrent protection.
The heater is controlled with a thermostat which does not have a marked "off" position. The branch circuit switch or circuit breaker is not "within sight from" the heater. For the branch circuit switch or circuit breaker to be permitted to serve as the disconnecting means for the heater, the switch or breaker must ___.

**(a) be readily accessible**
**(b) not control lamps or other appliances**
**(c) be capable of being locked in the open position**
**(d) be located within the dwelling unit or on the same floor as the heater**

50. Given: A metal underground water pipe is used as a grounding electrode and used to bond other electrodes together. The grounding electrode conductor is connected to the water pipe on the interior of the building. The connection of the grounding electrode conductor to the interior water pipe shall be made a maximum of ___ feet from the point where the water pipe enters the building.

**(a) 3 (b) 5 (c) 8 (d) 10**

# OPEN BOOK EXAM #3

## 50 QUESTIONS
## TIME LIMIT - 2 HOURS

**TIME SPENT** ☐ **MINUTES**

**SCORE** ☐ **%**

**JOURNEYMAN OPEN BOOK EXAM #3 Two Hour Time Limit**

1. Potential transformers, and other switchboard devices with potential coils shall be supplied by a circuit that is protected by standard overcurrent devices rated ____ amperes or less.

**(a) 15 (b) 20 (c) 25 (d) 30**

2. Which of the following is a false statement?

**(a) Where a building is supplied by more than one service, a permanent plaque or directory shall be installed at each service disconnect denoting the location of all other services.**
**(b) Service conductors supplying a building are permitted to pass through the interior of another building.**
**(c) Conductors other than service conductors shall not be installed in the same service raceway.**
**(d) Conductors run above the top level of a window shall be permitted to be less than 3 feet away from a window that is designed to be opened.**

3. Type ____ cable consists of three or more flat copper conductors placed edge-to-edge and separated and enclosed within an insulating assembly.

**(a) NMC (b) AC (c) MI (d) FCC**

4. A cord connector that is supported by a permanently installed cord pendant shall be considered ____.

**(a) receptacle outlet (b) permanent cord (c) lighting outlet (d) outlet device**

5. Equipment intended to break current at fault levels shall have an interrupting rating sufficient for the system voltage and the current which is ____ at the line terminals of the equipment.

**(a) at maximum (b) operating (c) available (d) required**

6. Electrodes of nonferrous metal shall be at least ____ in thickness.

**(a) 0.06mm (b) .186" (c) 1.52" (d) 0.06"**

7. Examples of resistance heaters are ____.

I. heating blankets II. heating tape III. heating barrel

**(a) I and II only (b) II and III only (c) III only (d) II only**

8. Metal surface raceways having splices and taps shall be permitted as long as the splices and taps and conductors do not fill the raceway more than ____ percent of the area of the raceway at that point.

**(a) 40 (b) 50 (c) 70 (d) 75**

9. Receptacles in damp or wet locations flush mounted shall ____.

I. be protected from the weather where located under canopies or marquees not subject to water run off
II. have an attachment plug cap inserted
III. be made weatherproof by means of a weatherproof faceplate assembly
IV. be located so that water accumulation is not likely to touch the outlet cover or plate

**(a) I only (b) II only (c) III only (d) I, II, III and IV**

10. Circuits that only supply neon tubing installations shall not be rated in excess of ____ amperes.

**(a) 15 (b) 20 (c) 30 (d) 50**

11. A portable motor which has an attachment plug and receptacle may use this type of attachment as the controller provided the motor does not exceed ____ hp.

**(a) 1/8 (b) 1/3 (c) 1 (d) 3**

12. Metal canopies supporting lampholders, shades, etc., exceeding ____ pounds shall not be less than 0.020 inch in thickness.

**(a) 4 (b) 6 (c) 8 (d) 10**

13. Live parts exposed on the front of a switchboard are present, the working space in front of the switchboard shall not be less than ____ inches.

**(a) 24 (b) 30 (c) 36 (d) 42**

14. Armored cable installed in thermal insulation shall have conductors rated at ____. The ampacity of cable installed in these applications shall be that of 60 degree C conductors.

**(a) 60 degrees C (b) 194 degrees F (c) 75 degrees C (d) 90 degrees F**

15. For hallways of ____ feet or more in length at least one receptacle outlet shall be required.

**(a) 6 (b) 8 (c) 10 (d) 12**

16. In panelboards, where the voltage on busbars is 150 volts and the bars are opposite polarity, held free in air, the minimum spacing between the parts is ____.

**(a) 3/4" (b) 1" (c) 1 1/2" (d) 2"**

17. Alkali-type battery cells in jars of conductive material shall be installed in trays of nonconductive material with not more than ____ 24 volt cells in the series circuit in any one tray.

**(a) ten (b) twenty (c) thirty (d) forty**

18. Exposed live parts within porcelain fixtures shall be suitably recessed and so located as to make it improbable that wires will come in contact with them. There shall be a spacing of at least ____ between live parts and the mounting plane of the fixture.

**(a) 1/4" (b) 1/8" (c) 1/2" (d) 3/4"**

19. The grounding conductor for secondary circuits of instrument transformers and for instrument cases shall not be smaller than #12 ____.

I. metal II. aluminum III. copper

**(a) I only (b) II only (c) III only (d) I, II or III**

20. A current-limiting overcurrent protective device is a device which will ____ the current flowing in the faulted circuit.

**(a) reduce (b) increase (c) maintain (d) none of these**

21. An office is to be wired with the number of receptacles unknown, the demand for the receptacles is ____ va per square foot.

**(a) 1 (b) 3 (c) 3.5 (d) 180**

22. In a recreational vehicle park with electrical supply, at least ___ % of the sites shall be equipped with 30 ampere, 125 volt receptacles.

**(a) 5 (b) 20 (c) 70 (d) 100**

23. No parts of pendants shall be located within a zone measured ____ feet horizontally and 8 feet vertically from the top of the bathtub rim.

**(a) 2 (b) 3 (c) 4 (d) 6**

24. The lead wires of heating cables are color coded for ____ identification.

**(a) lead (b) voltage (c) wire (d) cable**

25. Plug fuses must have what specific shape?

**(a) octagonal (b) square (c) hexagonal (d) round**

26. Fixtures in clothes closets shall be ____.

I. a surface-mounted or recessed incandescent fixture with a completely enclosed lamp
II. a surface-mounted or recessed fluorescent fixture
III. pendant fixture

**(a) I only (b) I and II only (c) I and III only (d) I, II and III**

27. All heating elements that are replaceable ____ and are a part of an electric heater shall be legibly marked with the rating in volts and watts, or in volts and amperes.

**(a) in the shop (b) by the manufacturer (c) in the field (d) none of these**

28. Plug fuses and fuseholders can be used in circuits supplied by a system having a grounded neutral and having no conductor at over ____ volts to ground.

**(a) 115 (b) 120 (c) 125 (d) 150**

29. EMT shall not be used ____.

**(a) for exposed work**
**(b) where protected from corrosion solely by enamel**
**(c) for concealed work**
**(d) none of these**

30. Where a motor is connected to a branch circuit by means of an attachment plug and receptacle and individual overload protection is omitted, the rating of the attachment plug and receptacle shall not exceed ____ or 250 volts.

**(a) 15 amperes at 110 volts**
**(b) 20 amperes at 115 volts**
**(c) 25 amperes at 120 volts**
**(d) 15 amperes at 125 volts**

31. All type FCC cable connections shall use connectors identified for their use, installed such that ____ against dampness and liquid spillage are provided.

I. electrical continuity II. insulation III. sealing

**(a) I only (b) II only (c) III only (d) I, II and III**

32. The disconnecting means of a hermetic-type refrigerator compressor shall have an ampacity of at least ____ of the nameplate full load current.

**(a) 125%   (b) 80%   (c) 100%   (d) 115%**

33. Fixtures shall be so constructed that adjacent combustible material will not be subject to temperatures in excess of ____ degrees C.

**(a) 60   (b) 75   (c) 90   (d) 110**

34. A factory installed duplex receptacle in a baseboard heater, where the heater is to be permanently installed in a commercial building is ____.

**(a) prohibited by the code**
**(b) allowed only when the receptacle is factory connected to the heater circuit**
**(c) not allowed to be used as the required receptacle outlet for flexible cords with attach ment plugs, when wired on a separate circuit from the heater circuit**
**(d) allowed to be used in lieu of the required receptacle outlet for flexible cords with attach ment plugs, when wired on a separate circuit from the heater circuit**

35. Type FCC cable, cable connectors, and insulating ends shall be covered with carpet squares no larger than ____ square.

**(a) 24"   (b) 914"   (c) 36mm   (d) 36"**

36. Vegetation such as trees shall not be used for support of ____ .

**(a) lighting fixtures   (b) brackets or clamps   (c) overhead conductor spans   (d) none of these**

37. Fixed electric space heating loads shall be computed at ____ percent of the total connected load; however in no case shall a feeder load current be less than the rating of the largest branch circuit supplied.

**(a) 80   (b) 100   (c) 115   (d) 125**

38. The adjustable speed drive incoming branch circuit or ____ to power conversion equipment included as a part of an adjustable speed drive system shall be based on the rated input to the power conversion equipment.

**(a) service   (b) feeder   (c) lateral   (d) none of these**

39. Separation of junction box from motor shall be permitted to be separated from the motor not more than ____.

**(a) 6 feet   (b) 4 feet   (c) 1.83   (d) none of these**

40. A single 1500w cord and plug connected load on 120v would draw ___ amps, this requires a number ____ wire and ____ circuit breaker for the branch circuit.

**(a) 8 - #14 - 15 amp (b) 10.5 - #14 - 15 amp (c) 12.5 - #14 - 15 amp (d) 12.5 - #12 - 20 amp**

41. SE cable used to supply ____ shall not be subject to conductor temperatures in excess of the temperature specified for the type of insulation involved.

**(a) lighting (b) appliances (c) motors (d) generators**

42. Torque motors are rated for operation ____.

**(a) at full torque (b) at F.L.C. (c) at standstill (d) with code letter**

43. The rating of an overcurrent device for a capacitor shall be ____.

**(a) not over 20 amp (b) as low as practicable**
**(c) less than 50 amp (d) none of these**

44. ____ of insulating material shall be permitted to be used without boxes in exposed cable wiring.

I. Switch devices II. Outlet devices III. Tap devices

**(a) I only (b) II only (c) III only (d) I, II and III**

45. The following pool equipment shall be grounded ____.

I. ground-fault circuit-interrupters
II. transformer enclosures
III. electric equipment located within 5 feet of the inside wall of the pool

**(a) III only (b) II and III only (c) II only (d) I, II and III**

46. It is the intent of the Code that ____ wiring or the construction of equipment need not be inspected at the time of installation of the equipment, if the equipment has been listed by a qualified electrical testing laboratory.

**(a) factory-installed internal (b) factory-installed**
**(c) underground (d) raceway**

47. Distances from signs, radio, and TV antennas, tanks or other nonbuilding or nonbridge structures, clearances, vertical, diagonal and horizontal, shall not be less than ____ feet.

**(a) 2 (b) 3 (c) 6 (d) 8**

48. Any motor application shall be considered as ____ unless the nature of the apparatus it drives is such that the motor will not operate continuously with load under any condition of use.

**(a) short-time duty** **(b) varying duty**
**(c) continuous duty** **(d) periodic duty**

49. An overcurrent trip unit of a circuit shall be connected in series with each ____.

**(a) ungrounded conductor** **(b) grounded conductor**
**(c) overcurrent device** **(d) transformer**

50. The grounded conductor of a mineral-insulated, metal-sheathed cable shall be identified at the time of installation by ____ marking at its termination.

**(a) distinctive (b) neutral (c) solid (d) identified**

# OPEN BOOK EXAM #4

## 50 QUESTIONS
## TIME LIMIT - 2 HOURS

**TIME SPENT** ☐ **MINUTES**

**SCORE** ☐ **%**

**JOURNEYMAN OPEN BOOK EXAM #4** **Two Hour Time Limit**

1. Where used as switches in 120 volt and 277 volt fluorescent lighting circuits, circuit breakers shall be marked ____.

**(a) UL (b) SWD (c) AMPS (d) VA**

2. The grounding electrode conductor shall be ____ and shall be installed in one continuous length without a splice or joint.

I. solid II. solid or stranded III. insulated, covered or bare

**(a) I only (b) I and III (c) II and III (d) III only**

3. The disconnecting means for motor circuits rated 600v, nominal, or less, shall have an ampere rating of what percent of the motor F.L.C.?

**(a) 100% (b) 125% (c) 115% (d) 140%**

4. Recessed portions of enclosures for flush recessed fixtures shall be spaced from combustible material by at least ____.

**(a) 1/4" (b) 3/4" (c) 1" (d) 1/2"**

5. Where it is impracticable to locate the service head above the point of attachment the service head location shall be permitted no further than how many feet from the point of attachment?

**(a) 1' (b) 2' (c) 3' (d) 4'**

6. For fixed multi-outlet assemblies where a number of appliances are likely to be used simultaneously, calculate a load of 180 volt-amps for each ____ ft.

**(a) 1 (b) 2 (c) 3 (d) 5**

7. Screw-type pressure terminals used with #14 or smaller copper conductors in motor controllers shall be torqued to a minimum of ____ pound-inches.

**(a) 7 (b) 10 (c) 12 (d) 20**

8. The identification of terminals to which a grounded conductor is to be connected shall be substantially ____ in color.

**(a) brass (b) copper (c) green (d) white**

9. Heating panels or panel sets, installed under floor covering, shall not exceed ____ watts per square foot of heated area.

**(a) 16 1/2 (b) 33 (c) 15 (d) 45**

10. The intent of the Code is to permit the exemption of receptacles which are located specifically for appliances such as ____ from GFCI protection for personnel.

I. hedge trimmers II. freezers III. refrigerators

**(a) I only (b) I and II only (c) II and III only (d) I, II and III**

11. Each electric appliance shall be provided with a nameplate, giving the identifying name and the rating in ____.

I. volts and watts II. watts and amps III. volts and amperes

**(a) I only (b) I or III (c) I or II (d) II or III**

12. Ground-fault protection of equipment shall be provided for solidly grounded wye electrical services of more than 150 volts to ground, but not exceeding 600 volts phase-to-phase for each service disconnecting means rated ____ amperes or more.

**(a) 200 (b) 600 (c) 800 (d) 1000**

13. For industrial establishments only, omission of overcurrent protection shall be permitted at points where busways are reduced in size, provided that the smaller busway does not extend more than ____ feet and has a current rating at least equal to ____ the rating or setting of the overcurrent device next back on the line.

**(a) 30' ... 80% (b) 50' ... 1/3 (c) 20' ... 1/2 (d) 40' ... 75%**

14. When conduit nipples having a maximum length not to exceed 24" are installed between boxes ____.

I. the nipple can be filled 75%
II. note 8 derating does apply
III. note 8 derating does not apply
IV. the nipple can be filled 60%

**(a) I and II (b) II and IV (c) III and IV (d) I and III**

15. Compliance with the provisions of the Code will result in ____.

**(a) good electrical service (b) an efficient system (c) freedom from hazard (d) all of these**

16. The total rating of a plug connected room air-conditioner where lighting units or other appliances are also supplied shall not exceed ____ percent.

**(a) 80 (b) 70 (c) 50 (d) 40**

17. What is the minimum number of overload units such as heaters, trip coils, or thermal cutouts allowed for a three-phase AC motor protection?

**(a) 1 (b) 2 (c) 3 (d) none of these**

18. All conductors the size below can be connected in parallel except ____.

**(a) #250 kcmil (b) #2/0 (c) #1 (d) #1/0**

19. Where raceways are exposed to widely different temperatures they shall be ____.

**(a) sealed (b) bonded (c) grounded (d) isolated**

20. When installing rigid nonmetallic conduit ____.

I. all joints shall be made by an approved method
II. there shall be support within 2 feet of each box, cabinet
III. all cut ends shall be trimmed inside and outside to remove rough edges

**(a) I, II and III (b) I and III (c) I and II (d) II and III**

21. The minimum size copper equipment grounding conductor required on a motor branch circuit with a 30 amp circuit breaker and #12 copper conductors is ____.

**(a) #10 (b) #8 (c) #12 (d) #14**

22. A raceway including the end fitting shall not use more than ____ inches into a panel containing 42 spaces for overcurrent devices.

**(a) 8 (b) 2 (c) 10 (d) 3**

23. Junction boxes for pool lighting shall not be located less than ____ feet from the inside wall of a pool unless separated by a fence or wall.

**(a) 3 (b) 4 (c) 6 (d) 8**

24. The unit lighting load for dwellings expressed in va per square foot is ____ va.

**(a) 2 (b) 5 (c) 3 (d) none of these**

25. Metal plugs or plates used with non-metallic boxes shall be recessed ____.

**(a) 3/8" (b) 1/2" (c) 1/4" (d) 1/8"**

26. Supplementary overcurrent devices shall ____.

**(a) not be required to be readily accessible**
**(b) be used as a substitute for branch-circuit overcurrent devices**
**(c) be readily accessible**
**(d) rated not over 15 amp**

27. Mats of insulating rubber or other suitable floor insulation shall be provided for the operator where the voltage to ground exceeds ____ on live-front switchboards.

**(a) 50 (b) 100 (c) 120 (d) 150**

28. A unit or assembly of units or sections, and associated fittings, forming a rigid structural system used to support cables and raceways would be the definition of ____.

**(a) wireway (b) multi-outlet assembly (c) cable tray (d) FCC**

29. A pliable raceway is a raceway which can be bent ____ with a reasonable force, but without other assistance.

**(a) with heat (b) without heat (c) by hand (d) easily**

30. What is the demand factor for five household clothes dryers?

**(a) 70% (b) 80% (c) 50% (d) 100%**

31. Non-current carrying metal parts of electrical equipment shall be kept how far from lightning rod conductors?

**(a) 3' (b) 6' (c) 8' (d) 10'**

32. Busways shall be securely supported, unless otherwise designed and marked at intervals not to exceed ____ feet.

**(a) 10 (b) 5 (c) 3 (d) 8**

33. Where it is unlikely that two dissimilar loads will be in use simultaneously, it shall be permissible to ____ of the two in computing the total load of a feeder.

**(a) omit both** **(b) omit the larger**
**(c) omit the smaller** **(d) omit neither**

34. Which of the following electrodes must be supplemented by an additional electrode?

**(a) metal underground water pipe** **(b) metal frame of a building**
**(c) ground ring** **(d) concrete encased**

35. In judging equipment, considerations such as the following shall be evaluated:

I. mechanical strength II. cost III. arcing effects IV. guarantee

**(a) I only (b) I and II (c) II and IV (d) I and III**

36. For the use of nonmetallic surface extensions the building _____.

I. cannot exceed three floors
II. is occupied for office purposes
III. is occupied for residential purposes

**(a) I only (b) II only (c) II and III (d) I, II and III**

37. When a flat cable assembly is installed less than ____ feet from the floor, it shall be protected by a metal cover identified for the use.

**(a) 8 (b) 10 (c) 12 (d) 15**

38. Pendant conductors longer than ____ shall be twisted together where not cabled in a listed assembly.

**(a) 12" (b) 18" (c) 2' (d) 3'**

39. Cablebus shall be permitted to be used for ____.

I. services II. feeders III. branch circuits

**(a) I only (b) II only (c) II and III (d) I, II and III**

40. Each vented cell shall be equipped with a ____ designed to prevent destruction of the cell.

**(a) gas arrestor (b) insulator (c) flame arrestor (d) electrolyte**

41. Thermoplastic insulation may stiffen at temperatures colder than minus ____ degrees C, requiring care be exercised during installation.

**(a) 5 (b) 10 (c) 15 (d) 30**

42. Flexible cords shall **not** be used in all but one of the following:

**(a) substitute for fixed wiring**
**(b) where run through holes in walls**
**(c) where attached to the building surface**
**(d) for pendants wiring fixtures, portable lamps, elevator cables**

43. The minimum ampacity for a 120/240v service entrance conductors is ____ amps.

**(a) 15 (b) 30 (c) 60 (d) 100**

44. A fixture that exceeds ____ inches in any dimension shall not be supported by the screw shell of a lampholder.

**(a) 8 (b) 10 (c) 12 (d) 16**

45. Lighting track which operates at 30 volts or higher shall be installed at least ___ feet above the finished floor.

**(a) 3 (b) 5 (c) 8 (d) 10**

46. Which of the following is the maximum number of current-carrying conductors that can be used at any cross-section of a wireway?

**(a) 100 (b) 30 (c) 50 (d) 40**

47. The following letter suffixes shall indicate the following:

____ -for two insulated conductors laid parallel within an outer nonmetallic covering.

**(a) D (b) M (c) R (d) N**

48. The means of identification of each system phase conductor, wherever accessible, may be by ____.

I. tagging, or other equally effective means
II. marking tape
III. separate color coding

**(a) I only (b) II only (c) III only (d) I, II or III**

49. For dwelling units, the computed floor area at 3va per square foot does NOT include ____.

I. bathrooms II. garages III. open porches

**(a) I and III only (b) II and III only (c) I and II only (d) I, II and III**

50. The screw shell contact of lampholders in grounded circuits shall be connected to the ____ conductor.

**(a) green** **(b) grounding**
**(c) ungrounded** **(d) grounded**

# OPEN BOOK EXAM #5

## 50 QUESTIONS
## TIME LIMIT - 2 HOURS

**TIME SPENT** ☐ **MINUTES**

**SCORE**  **%**

**JOURNEYMAN OPEN BOOK EXAM #5** **Two Hour Time Limit**

1. Ground-fault protection that functions to open the service disconnecting means ____ protect(s) service conductors or the service disconnecting means.

**(a) will (b) will not (c) adequately (d) totally**

2. Which of the following is a false statement?

**(a) direct buried conductors are required to be spliced in a splice box.**
**(b) direct buried conductors are permitted to be soldered.**
**(c) where wire connectors are used for splicing direct buried conductors, the connectors must be listed for such use.**
**(d) where necessary to prevent physical damage, direct buried conductors shall be protected by raceways, boards sleeves, or other approved means.**

3. The Code requires that heating panels be separated from outlet boxes that are to be used for mounting fixtures not less than ____ inches.

**(a) 12 (b) 8 (c) 6 (d) 10**

4. At least ____ inches of free conductor shall be left at each outlet and switch point.

**(a) 4 (b) 6 (c) 8 (d) 12**

5. It shall be permissible to apply a demand factor of 75% to the nameplate-rating load of 4 or more ____ fastened in place in a dwelling.

I. water heaters II. dishwashers III. clothes dryers

**(a) I only (b) II only (c) I and II only (d) I, II and III**

6. Where outdoor lampholders have terminals that puncture the insulation and make contact with the conductors, they shall be attached only to ____.

**(a) conductors with rubber insulation**
**(b) solid conductors**
**(c) conductors of the stranded type**
**(d) a #12 conductor**

7. Lamp tie wires, mounting screws, clips, and decorative bands on glass lamps spaced not less than ____ inches from lamp terminals shall not be required to be grounded.

**(a) 1 1/4 (b) 1 1/2 (c) 2 (d) 4**

8. Class II locations are those that are hazardous because of _____.

**(a) the presence of combustible dust**
**(b) over 8' depth of water**
**(c) flammable gases or vapors may be present in the air**
**(d) easily ignitible fibers are stored or handled**

9. Where conduit is threaded in the field, a standard conduit cutting die with a ____ inch taper per foot shall be used.

**(a) 1/2 (b) 3/4 (c) 1 (d) 1 1/4**

10. Equipment grounding conductors, when installed, ____ be included when calculating conduit fill.

**(a) should (b) shall (c) should not (d) shall never**

11. In a straight run of rigid nonmetallic conduit between securely mounted boxes, expansion joints are required where the computed length change due to thermal expansion or contraction is at least ___ inch or more.

**(a) 1/8 (b) 1/4 (c) 3/8 (d) 1/2**

12. The minimum feeder allowance for show window lighting expressed in volt-amps per linear foot shall be ____ va.

**(a) 100 (b) 200 (c) 300 (d) 180**

13. Angle pull dimensional requirements apply to junction boxes only when the size of conductor is equal to or larger than ____.

**(a) #0 (b) #4 (c) #3/0 (d) #6**

14. The maximum length of a bonding jumper on the outside of a raceway is ____.

**(a) 3' (b) 6' (c) 8' (d) none of these**

15. Rigid nonmetallic conduit may be used ____.

**(a) above ground in direct sunlight**
**(b) as a support for lighting fixtures**
**(c) as a grounding conductor**
**(d) all of these**

16. MI cable has _____.

**(a) solid copper conductors**
**(b) outer sheath to provide mechanical protection**
**(c) an adequate path for grounding purposes**
**(d) all of these**

17. Which of the following may be used as a feeder from the service equipment to a mobile home?

I. a permanently installed feeder   II. one 50 amp power supply cord

**(a) I only   (b) II only   (c) either I or II   (d) neither I nor II**

18. Multispeed motors shall be marked with the code letter designating the locked-rotor _____ per horsepower for the highest speed at which the motor can be started.

**(a) amps   (b) F.L.C.   (c) kva   (d) watts**

19. The length of a type S cord connecting a trash compactor must not exceed _____.

**(a) 18"   (b) 4'   (c) 36"   (d) 2'**

20. Electrical installations in hollow spaces, vertical shafts and ventilation or air-handling ducts shall be so made that the possible spread of fire or products of combustion will not be _____.

**(a) substantially increased   (b) allowed   (c) exposed   (d) under rated**

21. Electric equipment shall be installed in a neat and _____ manner.

**(a) efficient   (b) safe   (c) workmanlike   (d) orderly**

22. The space measured horizontally above a show window must have at least one receptacle for each _____ linear feet.

**(a) 12   (b) 10   (c) 8   (d) 6**

23. Conductor overload protection is not required if _____.

**(a) conductors are oversized by 125%**
**(b) conductors are part of a limited-energy circuit**
**(c) interruption of the circuit can create a hazard**
**(d) none of these**

24. The distance between a cable or conductor entry and its exit from the box shall be not less than ____ times the outside diameter, over sheath, of that cable or conductor, 1000 volt system.

**(a) 6 (b) 18 (c) 36 (d) 48**

25. A thermal barrier shall be required if the space between the resistors and reactors and any combustible material is less than ____ inches.

**(a) 4 (b) 6 (c) 8 (d) 12**

26. An attachment plug connecting to a receptacle shall ____ the equipment grounding conductor.

**(a) have conductors the same size as**
**(b) provide for first-make, last-break of**
**(c) provide a twist-lock connection for**
**(d) none of these**

27. When more than one calculated or tabulated ampacity could apply for a given circuit length, the ____ value shall be used.

**(a) lowest (b) average (c) highest (d) none of these**

28. Cable splices made and insulated by approved methods shall be permitted within a cable tray provided they are accessible and ____.

**(a) have a hinged cover**
**(b) are crimped properly**
**(c) are not over 600 volt**
**(d) do not project above the side rails**

29. Electronically actuated fuses may or may not operate in a current limiting fashion, depending on the ____.

**(a) ambient temperature (b) type of control selected**
**(c) listing (d) torque**

30. Connection by means of wire binding screws or studs and nuts having upturned lugs or equivalent shall be permitted for ____ or smaller conductors.

**(a) #10 (b) #8 (c) #6 (d) none of these**

31. Electrical nonmetallic tubing is permitted to be used in sizes up to ____.

**(a) 1" (b) 2" (c) 3" (d) 4"**

32. Ampacity of fixture wire is determined ____.

**(a) by referring to the ampacity Table 310-16**
**(b) by calculation, using the expected temperature rise of the fixture**
**(c) from a table in article 402 of the Code**
**(d) none of these**

33. Pull-type canopy switches shall not be located more than ____ from the center of the canopy.

**(a) 1 1/2"   (b) 2"   (c) 3"   (d) 3 1/2"**

34. Means shall be provided to ensure that the ____ is energized when the first heater circuit is energized.

**(a) ballast   (b) fan circuit   (c) coil   (d) relay**

35. A pool recirculating pump motor receptacle shall be permitted not less than ____ feet from the inside walls of the pool.

**(a) 5   (b) 8   (c) 10   (d) 15**

36. Fixtures shall be wired with conductors having insulation suitable for ____ to which the conductors will be subjected.

I. environmental conditions   II. current-voltage   III. temperature

**(a) II only   (b) III only   (c) I, II and III   (d) II and III**

37. What is the minimum working clearance on a circuit 120 volts to ground, exposed live parts on one side and no live or grounded parts on the other side of the working space?

**(a) 3'   (b) 3 1/2'   (c) 4'   (d) 6'**

38. The maximum weight of a light fixture that may be mounted on the screw shell of a brass socket is ____ pound(s).

**(a) 1/2   (b) 1   (c) 6   (d) none of these**

39. The grounded service conductor shall not be smaller than the required ____.

**(a) grounding electrode conductor**
**(b) largest phase conductor**
**(c) ungrounded service conductor**
**(d) largest equipment conductor**

40. Type UF cable shall be permitted for interior wiring in ____ locations.

I. dry II. wet III. corrosive

**(a) I only (b) I or II (c) I or III (d) I, II or III**

41. Type ____, a flat cable assembly, is an assembly of parallel conductors formed integrally with an insulating material web specifically designed for field installation in surface metal raceway.

**(a) FCC (b) FC (c) TC (d) SNM**

42. For feeder and service calculations a maximum of ____ of lighting track or fraction thereof shall be considered 150va.

**(a) 2' (b) 4' (c) 5' (d) 8'**

43. Under the optional method of calculation "other loads" are permitted a demand factor from Table 220-30, the first 10 kva of "other load" @ 100% and the remainder of "other load" at 40%. "Other load" could consist of which of the following?

I. electric heat II. electric range III. air conditioning

**(a) I only (b) II only (c) III only (d) I, II and III**

44. Reasonable efficiency of operation can be provided when ____ is taken into consideration in sizing the service-lateral conductors.

**(a) mechanical strength (b) ambient temperature (c) voltage drop (d) none of these**

45. Voltage shall not exceed 600 volts between conductors on branch circuits supplying only ballasts for electric-discharge lamps in tunnels with a height of not less than ____ feet.

**(a) 12 (b) 15 (c) 18 (d) 22**

46. Conduit encased in a concrete trench is considered a ____ location.

**(a) wet (b) dry (c) damp (d) moist**

47. The conductor between a lightning arrester and the line for installations operating at 1000 volts or more must be at least ____.

**(a) #14 copper (b) #6 copper (c) #8 copper (d) none of these**

48. What is the nominal battery voltage for an alkali type battery per cell?

**(a) 2.0 volt (b) 6.0 volt (c) 1.5 volt (d) 1.2 volt**

49. The conductors and equipment required or permitted by this Code shall be acceptable only if _____.

**(a) approved (b) identified (c) labeled (d) listed**

50. Where multiple rod, pipe, or plate electrodes are installed they shall be not less than _____ apart.

**(a) 18" (b) 6' (c) 8' (d) 10'**

# OPEN BOOK EXAM #6

## 50 QUESTIONS
## TIME LIMIT - 2 HOURS

**TIME SPENT** ☐ **MINUTES**

**SCORE** ☐ **%**

## JOURNEYMAN OPEN BOOK EXAM #6 Two Hour Time Limit

1. Where extensive metal in or on buildings may become energized and is subject to personal contact ____ will provide additional safety.

**(a) adequate bonding and grounding** **(b) bonding**
**(c) suitable ground detectors** **(d) none of these**

2. Single conductor cables shall be ____ or larger and shall be of a type listed for use in cable trays.

**(a) #1 (b) #1/0 (c) #4/0 (d) #250 kcmil**

3. The grounded conductor, when insulated, shall have insulation ____.

I. rated not less than 300 volts for solidly grounded neutral systems of 1 kv and over as described in section 250-184
II. which is suitable, other than color, for any ungrounded conductor of the same circuit on circuits of less than 1000 volts

**(a) I only (b) II only (c) either I or II (d) neither I nor II**

4. Which of the following is **not** true regarding rigid nonmetallic conduit?

**(a) extreme cold may cause some nonmetallic conduits to become brittle and therefore more susceptible to damage from physical contact**
**(b) can be used to support fixtures**
**(c) all cut ends shall be trimmed inside and outside to remove rough edges**
**(d) expansion joints shall be provided to compensate for thermal expansion and contraction**

5. Lighting track conductors shall be a minimum ____ AWG or equal, and shall be copper.

**(a) #16 (b) #14 (c) #12 (d) #10**

6. In a residence a multiwire branch circuit supplying more than one device or equipment on the same ____ shall be provided with a means to disconnect simultaneously all the hot conductors at the panelboard where the branch circuit originated.

**(a) branch-circuit (b) yoke (c) device (d) outlet assembly**

7. Cablebus shall be installed only for ____ work.

**(a) exposed (b) commercial (c) concealed (d) hazardous**

8. Knife switches rated for more than 1200 amperes at 250 volts ____.

**(a) are used only as isolating switches**
**(b) should be placed so that gravity tends to close them**
**(c) should be opened slowly under load**
**(d) should be connected so blades are not dead in open position**

9. A transverse metal raceway for electrical conductors, furnishing access to predetermined cells of a precast cellular concrete floor, which permits installation of conductors from a distribution center to the floor cells is called ____.

**(a) an underfloor raceway** **(b) a header duct**
**(c) a cellular raceway** **(d) a mandrel**

10. Because aluminum is not a magnetic metal, there will be no heating due to ____.

**(a) electrolysis (b) hysteresis (c) hermetic (d) galvanic action**

11. Fixtures shall be so constructed, or installed, or equipped with shades or guards that combustible material will not be subjected to temperatures in excess of ____.

**(a) 90°F (b) 86°F (c) 30°C (d) 90°C**

12. The ampacity of the phase conductors from generator terminals to the first overcurrent device shall not be less than ____ percent of the nameplate current rating of the generator.

**(a) 80 (b) 115 (c) 125 (d) 150**

13. All cut ends of rigid conduit shall be ____.

**(a) threaded (b) electrically continuous (c) reamed (d) cut square**

14. What size conductor shall be connected between the ground grid and all metal parts of swimming pools?

**(a) #8 (b) #10 (c) #6 (d) #4**

15. Exposed runs of armored cable shall closely follow the surface of the building or of running boards except lengths of not more than ____ inches at terminals where flexibility is necessary.

**(a) 24 (b) 30 (c) 36 (d) 48**

16. A cabinet or cutout box if constructed of sheet steel, the metal thickness shall not be less than ____ inch uncoated.

**(a) 0.053 (b) 0.503 (c) 0.040 (d) 0.373**

17. The minimum headroom of working spaces about control centers shall be ____.

**(a) 3' 6" (b) 5" (c) 6' 4" (d) 6' 6"**

18. Conductors of AC or DC circuits rated 600 volt or less, shall be permitted to occupy the same conduit if ____.

**(a) all conductors shall have an insulation voltage rating equal to the maximum circuit voltage rating of any conductor in the conduit**
**(b) all conductors shall have a 600 volt insulation rating**
**(c) conductors must have a dividing barrier in the raceway**
**(d) AC and DC are not permitted in the same raceway**

19. Where the service disconnecting means does not ____ the grounded conductor from the premises wiring, other means shall be provided for this purpose in the service equipment.

**(a) shut off (b) trip (c) isolate (d) disconnect**

20. Outlets for specific appliances such as laundry equipment, shall be within ____ feet of the appliance.

**(a) 4 (b) 6 (c) 8 (d) 10**

21. Which of the following is true?

**(a) the loads of outlets serving switchboards and switching frames in telephone exchanges shall be counted in branch-circuit computations**
**(b) a multiple receptacle shall be considered at not less than 420va for computations of other outlets**
**(c) the minimum general lighting load for a restaurant is 3 va per sq.ft.**
**(d) an electric clock may be connected to a small appliance branch circuit**

22. Cable or raceway that is installed through bored holes in wood members, holes shall be bored so that the edge of the hole is not less than 1 1/4" from the nearest edge of the wood member. Where this distance cannot be maintained the cable or raceway shall be protected from penetration by nails and screws by a steel plate or bushing, at least ____ inch thick, and of appropriate length and width installed to cover the area of the wiring.

**(a) 1/16 (b) 1/8 (c) 3/16 (d) 1/4**

23. Except where fire stops are required, it shall be permissible to extend cablebus vertically through dry floors and platforms, provided the cablebus is totally enclosed at the point where it passes through the floor or platform and for a distance of ____ feet above the floor or platform.

**(a) 6 (b) 8 (c) 10 (d) 4**

24. Minimum headroom shall be provided for all working spaces about service equipment, switchboards, panelboards, or motor control centers except in service equipment or panelboards in dwelling units that do not exceed ____ amperes.

**(a) 150 (b) 200 (c) 175 (d) 300**

25. Fixtures which require aiming or adjusting after installation shall not be required to be equipped with an attachment plug or cord connector provided the exposed cord is ____.

I. not longer than that required for maximum adjustment
II. hard usage or extra-hard usage type

**(a) I only (b) II only (c) both I and II (d) neither I nor II**

26. ____ or larger conductors supported on solid knobs shall be securely tied thereto by tie wires having an insulation equivalent to that of the conductor.

**(a) #12 (b) #10 (c) #8 (d) #6**

27. ____ is defined as the shortest distance measured between a point on the top surface of any direct buried conductor, cable, conduit, or other raceway and the top surface of finished grade.

**(a) Depth (b) Cover (c) Gap (d) Soil**

28. Electric vehicle cable type EVJ ____.

I. comes in sizes #18-#500 kcmil II. is for extra hard usuage III. has thermoset insulation

**(a) I only (b) II only (c) III only (d) I, II and III**

29. Which of the following statements about MI cable is correct?

**(a) it may be used in any hazardous location**
**(b) it may be mounted flush on a wall in a wet location**
**(c) it shall be supported every 10 feet**
**(d) a single run of cable shall not contain more than four quarter bends**

30. Tap conductors in a metal raceway for recessed fixture connections shall be limited to ____ feet in length.

**(a) 2 (b) 4 (c) 6 (d) 10**

31. Where a permanent barrier is installed in a pull box, each section is considered as _____.

**(a) permanent barriers are not allowed (b) a separate box**
**(c) 60% of the box (d) the same box**

32. Two one ohm resistors in parallel, total resistance is ____ ohm.

**(a) 1 (b) 2 (c) 1/2 (d) cannot be calculated**

33. Underground service conductors carried up a pole must be protected from mechanical injury to a height of at least ____ feet.

**(a) 12 (b) 8 (c) 15 (d) 9**

34. In straight pulls, the length of the box shall be not less than ____ times the trade diameter of the largest raceway.

**(a) 4 (b) 6 (c) 8 (d) 12**

35. Wall-mounted ovens and counter-mounted cooking units complete with provisions for mounting and for making electrical connections, shall be permitted to be ____.

I. plug and cord connected II. permanently connected

**(a) I only (b) II only (c) either I or II (d) neither I nor II**

36. Receptacles connected to circuits having different ____ on the same premises shall be of such design that the attachment plugs used on these circuits are not interchangeable.

I. current (AC or DC) II. frequencies III. voltages IV. wattages

**(a) I and III only (b) I and II only (c) I, II and III only (d) I, II, III and IV**

37. Five pieces of kitchen equipment in a restaurant would have a feeder demand factor of ____ percent.

**(a) 65 (b) 70 (c) 80 (d) 90**

38. Which of the following is **not** true?

**(a) A demand factor from Table 220-19 could be applied to a household counter-mounted cooking unit of 1760 watts.**
**(b) Ten household clothes dryers have a demand factor of 50%.**
**(c) A demand factor from Table 220-19 could be applied to a 1 3/4 kw wall-mounted oven.**
**(d) Table 220-19 is permitted for a branch circuit to a household range.**

39. Where the service overcurrent devices are locked or sealed, or otherwise not readily accessible, branch-circuit overcurrent devices shall be ____.

I. of lower ampere rating than the service overcurrent device
II. mounted in an readily accessible location
III. installed on the load side

**(a) I only (b) II only (c) III only (d) I, II and III**

40. Grounding conductors and bonding jumpers shall be connected by ____ or other listed means.

I. listed clamps II. listed pressure connectors III. exothermic welding

**(a) I only (b) II only (c) III only (d) I, II or III**

41. Cable trays shall ____.

I. have side rails or equivalent structural members
II. not present sharp edges or burrs
III. have suitable strength and rigidity

**(a) I only (b) I and II only (c) III only (d) I, II and III**

42. A raceway containing 30 current carrying conductors, the ampacity of each conductor shall be reduced ____ percent.

**(a) 80 (b) 70 (c) 45 (d) 50**

43. The Code requires all conductors that attach to a cablebus to be in the same raceway because ____.

**(a) of less voltage drop (b) the cost is less (c) it is easier to service (d) of inductive current**

44. What is the minimum size conductor that may be used for an overhead feeder which is 35 feet in length from a residence to a remote garage?

**(a) #10 cu (b) #8 cu (c) #6 cu (d) #4 cu**

45. Nonmetallic sheath cable must be supported within ____ of a metal box.

**(a) 6" (b) 12" (c) 24" (d) 48"**

46. The temperature limitation of MI cable is based on the ____.

**(a) ambient temperature**
**(b) conductor insulation**
**(c) insulating materials used in the end seal**
**(d) none of these**

47. All electric equipment, including power supply cords used with storable swimming pools shall be protected by ____.

**(a) GFCI (b) fuses (c) circuit breakers (d) current limiting fuses**

48. Service conductors shall be attached to the disconnecting means by pressure connectors, clamps or other approved means, except connections that depend on ____ shall not be used.

**(a) solder (b) tension (c) bolts (d) pressure**

49. Which of the following wiring methods is permitted through an air conditioning duct?

**(a) electrical metallic tubing**
**(b) PVC**
**(c) no wiring method is permitted in an A/C duct**
**(d) romex**

50. Conductors run above the top level of a window shall be permitted to be less than the ____ requirement for clearance from a window.

**(a) 2' (b) 3' (c) 4' (d) 8'**

# OPEN BOOK EXAM #7

## 50 QUESTIONS
## TIME LIMIT - 2 HOURS

**TIME SPENT** ☐ **MINUTES**

**SCORE**  **%**

## JOURNEYMAN OPEN BOOK EXAM #7 Two Hour Time Limit

1. In general, switches shall be so wired that all switching is done in the ____ conductor.

**(a) grounded (b) ungrounded (c) both (a) and (b) (d) neither (a) nor (b)**

2. Material identified by the subscript letter ____ includes text extracted from other NFPA documents.

**(a) W (b) X (c) Y (d) Z**

3. Insulated conductors smaller than ____, intended for use as grounded conductors of circuits, shall have an outer identification of white or gray color.

**(a) #4 (b) #2 (c) #1/0 (d) #250 kcmil**

4. "Z.P." is an abbreviated marking used for motors to indicate ____.

**(a) single-phase (b) induction-protected**
**(c) thermally protected (d) impedance protected**

5. A pool panelboard, not part of the service equipment, shall have a grounding conductor installed between ____.

**(a) its grounding terminal and a separate ground**
**(b) its grounding terminal and a ground rod**
**(c) its grounding terminal and the grounding terminal of the service equipment**
**(d) its grounding terminal and bonding grid**

6. Overcurrent protective devices shall be so selected and coordinated as to permit the circuit protective devices used to clear a fault without the occurrence of extensive damage to the electrical components of the circuit. This fault shall be assumed to be ____.

I. between any circuit conductor and the grounding conductor or enclosing metal raceway
II. between two or more of the circuit conductors

**(a) I only (b) II only (c) both I and II (d) neither I nor II**

7. The ampacity for conductors is derated when the ambient temperature exceeds:

**(a) 30 degrees F (b) 72 degrees F (c) 86 degrees F (d) 104 degrees F**

8. Transformers isulated with a dielectric fluid installed indoors and rated over _____ shall be installed in a vault.

**(a) 112 1/2 kva (b) 35,000 va (c) 35 kv (d) 35 kva**

9. Which of the following requires a moisture seal at all points of termination?

**(a) underplaster extensions**
**(b) bare conductor feeders**
**(c) liquidtight flexible metal conduit**
**(d) mineral-insulated cable**

10. For a feeder supplying household cooking equipment and electric clothes dryers the maximum unbalanced load on the neutral conductor shall be considered as ______ of the load on the ungrounded conductors.

**(a) 40% (b) 50% (c) 70% (d) 80%**

11. Formal interpretations of the Code may be found in the _____.

**(a) National Electrical Code Handbook**
**(b) OSHA Standards**
**(c) NFPA Regulations Governing Committe Projects**
**(d) Life and Safety Handbook**

12. Sign lighting system equipment shall be at least _____ feet above areas accessible to vehicles unless protected from physical damage.

**(a) 14 (b) 15 (c) 18 (d) 22**

13. Where a transformer or other device is used to obtain a reduced voltage for the motor control circuit and is located in the controller, such transformer or other device shall be connected _____ for the motor control circuit.

I. to the load side of the disconnecting means
II. to the line side of the disconnecting means

**(a) I only (b) II only (c) either I or II (d) neither I nor II**

14. A _____ is a protective device for limiting surge voltages by discharging or bypassing surge current, and it also prevents continued flow of follow current while remaining capable of repeating these functions.

**(a) surge arrester (b) auto fuse (c) fuse (d) circuit breaker**

15. Type FCC cable shall be clearly and durably marked with ____.

I. material of conductors II. maximum temperature rating III. ampacity

**(a) I only (b) II only (c) III only (d) I, II and III**

16. No swimming pool lighting fixtures shall be installed for operation on supply circuits over ____ volts between conductors.

**(a) 24 (b) 50 (c) 120 (d) 150**

17. Only wiring methods recognized as ____ are included in the Code.

**(a) approved (b) suitable (c) listed (d) identified**

18. Service conductors between the street main and the first point of connection to the service entrance run underground is known as the service ____.

**(a) drop (b) loop (c) lateral (d) cable**

19. EMT installed in a wet location, shall have its coupling and connectors ____.

**(a) protected against corrosion (b) corrosion resistant**
**(c) raintight type (d) none of these**

20. Dual-voltage motors that have a different locked-rotor kva per horsepower on the two voltages shall be marked with the code letter for the voltage giving the ____ locked-rotor kva per horsepower.

**(a) highest (b) average (c) lowest (d) normal**

21. The Code requires in a dwelling a minimum of ____.

I. 3 volt-amps per square foot II. one 8 kw range
III. two small appliance circuits IV. one laundry circuit

**(a) I and II only (b) I, II and III only (c) I, III and IV only (d) I, II III and IV**

22. Outdoor electrical installations over 600 volts that are open to unqualified persons shall comply with ____.

**(a) Chapter 9 (b) Article 225 (c) Chapter 7 (d) Article 110**

23. The optional method of calculation is permitted for a multifamily dwelling if ____.

I. each dwelling unit is equipped with either electric space heating or air conditioning or both
II. no dwelling unit is supplied by more than one feeder

**(a) I only (b) II only (c) both I and II (d) neither I nor II**

24. Messenger supported wiring shall not be used ____.

I. where subject to severe physical damage
II. in hoistways

**(a) I only (b) II only (c) both I and II (d) neither I nor II**

25. Receptacles installed on ____ ampere branch circuits, shall be of the grounding type.

**(a) 15 and 20 (b) 25 (c) 30 (d) 40**

26. Class I locations are those that are hazardous because of ____.

**(a) the presence of combustible dust**
**(b) over 8' depth of water**
**(c) flammable gases or vapors are or may be present in the air**
**(d) the presence of easily ignitible fibers or flyings**

27. Which of the following about the equipment grounding conductor is/are true?

I. does not count as a current-carrying conductor
II. bare, covered or insulated shall be permitted
III. count one for each grounding conductor in conduit fill

**(a) I only (b) II and III only (c) I and III only (d) I, II and III**

28. Metal faceplates for devices shall be of ferrous metal not less than ____ inches in thickness.

**(a) 0.300 (b) 0.003 (c) 0.030 (d) none of these**

29. When a controller is **not** within sight from the motor location, the disconnect shall be capable of being ____ in the open position.

**(a) down (b) up (c) locked (d) shut-off**

30. A green wire with yellow stripes used in a branch-circuit would be the ____ conductor.

**(a) grounded (b) grounding (c) neutral (d) ungrounded**

31. Heaters installed within ____ feet of the outlet of an air-moving device, heat pump, A/C, elbows, baffle plates, or other obstructions in duct work may require turning vanes, pressure plates, or other devices on the inlet side of the duct heater to assure an even distribution of air over the face of the heater.

**(a) 2 (b) 3 (c) 4 (d) 6**

32. In a dwelling, a 20 ampere rated living room branch circuit can be loaded to a maximum of ____ amperes.

**(a) 10 (b) 15 (c) 16 (d) 20**

33. Conductor A.W.G. numbers vary ____ to the ampacity.

**(a) inversely (b) proportionally (c) directly (d) bi-laterally**

34. No receptacle shall be installed within ____ feet of the inside walls of a pool.

**(a) 10 (b) 15 (c) 18 (d) 20**

35. Electrically heated smoothing irons shall be equipped with an identified ____ means.

**(a) disconnecting (b) temperature-limiting (c) cooling (d) shut-off**

36. Type TC power and control cable may be used ____.

**(a) in outdoor locations when supported by a messenger cable**
**(b) as open cable on brackets**
**(c) where exposed to physical damage**
**(d) none of these**

37. Heavy-duty lamps are used on ____ ampere or larger circuits.

**(a) 15 (b) 20 (c) 25 (d) 30**

38. A switch box installed in a tiled wall may be recessed ____ behind the finished wall.

**(a) 1/4" (b) 3/8" (c) 1/2" (d) not at all**

39. Raceways on the outside of buildings shall be ____.

**(a) watertight and arranged to drain (b) weatherproof and covered**
**(c) raintight and arranged to drain (d) rainproof and guarded**

40. A new building will have two service heads, serviced by one service drop. What is the maximum distance apart that the Code permits the service heads to be located?

**(a) 36" (b) 48" (c) 6 feet (d) no maximum as long as the conductors will reach**

41. What is the area of square inches for a #8 bare conductor in a raceway?

**(a) 0.013 (b) 0.017 (c) 0.778 (d) 0.809**

42. Receptacles mounted on ____ need not be grounded.

**(a) outdoor circuits (b) garage walls**
**(c) portable generators (d) electric ranges**

43. Splices and taps shall not be located within fixture ____.

**(a) splice boxes (b) arms or stems (c) pancake boxes (d) none of these**

44. Floor boxes shall be considered to meet the requirements of the spacing of receptacles on walls if they are within ____ to the wall.

**(a) 18" (b) 20" (c) 24" (d) 30"**

45. ____ may be conected ahead of service switches.

I. Surge arrestors II. Current-limiting devices

**(a) I only (b) II only (c) neither I nor II (d) both I and II**

46. Which of the following may **not** be used in damp or wet locations?

**(a) AC armored cable (b) EMT (c) open wiring (d) rigid steel conduit**

47. Except where computations result in a major fraction of an ampere ____, such fractions may be dropped.

**(a) larger than 0.5 (b) 0.5 or larger**
**(c) smaller than 0.5 (d) 0.8 or larger**

48. In a dwelling it shall be permissible to apply a demand factor of ____ percent to the nameplate rating load of four or more appliances fastened in place.

**(a) 60 (b) 70 (c) 75 (d) 80**

49. The ampacity of a #250 kcmil IGS cable is ____ amperes.

**(a) 119 (b) 168 (c) 215 (d) 255**

50. Enclosures supported by suspended ceiling systems shall be fastened to the framing member by mechanical means such as ____.

I. clips identified for use II. screws III. rivets IV. bolts

**(a) I only (b) II only (c) II and IV only (d) I, II, III and IV**

# OPEN BOOK EXAM #8

## 50 QUESTIONS
## TIME LIMIT - 2 HOURS

**TIME SPENT** ☐ **MINUTES**

**SCORE** ☐ **%**

## JOURNEYMAN OPEN BOOK EXAM #8 Two Hour Time Limit

1. Where the number of current-carrying conductors in a raceway is seven, the individual ampacity of each conductor shall be reduced ____.

**(a) to 70% due to the number of conductors**
**(b) to 80% if they are continuous loads**
**(c) to both (a) and (b) if both conditions exist**
**(d) neither apply if the ambient temperature is below 30° C or 86° F**

2. Insulated bushings are required on conduit entering boxes, gutters, etc. if the conduit contains conductors as large as ____.

**(a) #2 (b) #4 (c) #0 (d) #6**

3. Plug-in-type overcurrent protection devices or plug-in-type main lug assemblies that are ____ shall be secured in place by an additional fastener that requires other than a pull to release the device from the mounting means on the panel.

**(a) three-phase only (b) 480v (c) back fed (d) none of these**

4. Fluorescent lighting fixtures may be used as raceways if ____.

**(a) they are connected by a conduit wiring method**
**(b) they are wired so that conductors are not closer than 3" from the ballast**
**(c) listed for use as a raceway**
**(d) none of these**

5. When supplying a nominal 120v rated air-conditioner, the length of the flexible supply cord shall not exceed ____ feet.

**(a) 4 (b) 6 (c) 8 (d) 10**

6. Which of the following is the maximum allowable rating of a permanently connected appliance where the branch circuit overcurrent device is used as the appliance disconnecting means?

**(a) 1/8 hp (b) 1/4 hp (c) 1/2 hp (d) 1 hp**

7. The number of #12 THW conductors allowed in a 3/4" IMC conduit will be ____ the number of #12 TW conductors allowed in a 3/4" conduit.

**(a) equal to (b) greater than (c) less than (d) none of these**

8. When connections are made in the white wire in a multiwire circuit at receptacles, they are required to be made _____.

**(a) connected to the silver terminal on the duplex**
**(b) to the brass colored terminal**
**(c) with a pigtail to the silver terminal**
**(d) none of these**

9. Unguarded live parts above working space shall be maintained at an elevation of ____ for 4160 volts.

**(a) 8' (b) 8' 6" (c) 9' (d) 10'**

10. Which of the following is **not** true?

**(a) the receptacle outlet spacing in a motel room can be more than 12' from outlet to outlet**
**(b) a two-family dwelling requires at least one receptacle outlet outdoors for each dwelling unit at grade level**
**(c) a vehicle door in an attached garage is not considered as an outdoor entrance**
**(d) a vehicle door in an attached garage is considered as an outdoor entrance**

11. Service-drop conductors shall have _____.

I. adequate mechanical strength
II. sufficient ampacity to carry the load as computed in accordance with Article 220

**(a) I only (b) II only (c) both I and II (d) neither I nor II**

12. The frame of a clothes dryer shall be permitted to be grounded to the grounded circuit conductor if _____.

I. the grounded conductor is insulated
II. the grounded conductor is not smaller than #10 copper
III. the supply circuit is 120/240v single-phase

**(a) I only (b) II only (c) III only (d) I, II and III**

13. A 20 ampere rated branch circuit serves four receptacles. The rating of the receptacles must not be less than ____ amperes.

**(a) 20 (b) 15 (c) 25 (d) none of these**

14. When the voltage to a building is 480/277, and the service drop runs not more than four feet past the edge of the overhang of the roof, how high must it be above the roof?

**(a) 18" (b) 3' (c) 4' (d) 8'**

15. The ampacity of a feeder conductor supplying two or more 2-wire branch circuits shall not be less than ____ amps.

**(a) 20 (b) 25 (c) 30 (d) 40**

16. Where the calculated number of conductors, all of the same size, includes a decimal fraction, the next higher whole number shall be used if ____.

**(a) .5 and larger (b) .6 and larger (c) .7 and larger (d) .8 and larger**

17. The height of a circuit breaker used as a switch shall not exceed ____ above the floor.

**(a) 4' (b) 4 1/2' (c) 5' (d) 6' 7"**

18. The number of #12 conductors permitted in a 3" x 2" x 1 1/2" deep device box is ____.

**(a) 6 (b) 5 (c) 4 (d) 3**

19. What is the minimum height of a service drop attachment to a building?

**(a) 8 feet (b) 10 feet (c) 12 feet (d) 15 feet**

20. Heating cables shall be furnished with nonheating leads at least ____ in length.

**(a) 7' (b) 8' (c) 10' (d) 12'**

21. In a dwelling, which appliance shall be grounded?

**(a) toaster (b) can opener (c) blender (d) aquarium**

22. #0 copper conductors in vertical raceway shall be supported at intervals not exceeding ____ feet.

**(a) 50 (b) 75 (c) 100 (d) 125**

23. Rigid conduit buried in an area subject to heavy vehicular traffic shall have a minimum cover of ____ inches.

**(a) 6 (b) 12 (c) 18 (d) 24**

24. A single-family dwelling contains a 200 amp single-phase service panel supplied with #2/0 THW conductors. The minimum size bonding jumper for this service is ____.

**(a) #6 aluminum (b) #6 copper (c) #4 aluminum (d) #4 copper**

25. A 1 1/2" rigid metal nipple with three conductors can be filled to an area of ____ square inches.

**(a) .98 (b) 1.07 (c) 1.2426 (d) 1.34**

26. Grounding electrode conductors smaller than #6 shall be in ____.

I. EMT II. IMC III. rigid PVC IV. rigid metal conduit

**(a) I and IV only (b) I, II and IV only (c) II and IV only (d) I, II, III and IV**

27. The nominal gas pressure for IGS cable insulation shall be ____ pounds per square inch gage.

**(a) 5 (b) 10 (c) 15 (d) 20**

28. Type SE service-entrance cables shall be permitted in interior wiring systems where all of the circuit conductors of the cable are of the ____ type.

I. rubber-covered II. thermoplastic III. metal

**(a) I and II only (b) II only (c) II and III only (d) I, II and III**

29. Elevator traveling cables for operating ____ circuits shall contain nonmetallic fillers as necessary to maintain concentricity.

I. signal II. control

**(a) I only (b) II only (c) both I and II (d) neither I nor II**

30. Where installed in a metal raceway all conductors of all feeders using a common neutral shall be ____.

**(a) insulated for 600 volt (b) enclosed within the same raceway**
**(c) shielded (d) none of these**

31. For household ranges rated ___ or more rating, the minimum branch circuit rating shall be 40 amperes.

**(a) 4 kw (b) 6 kw (c) 8 kw (d) 8 3/4 kw**

32. Receptacles located within ____ feet of the inside walls of a pool shall be protected by a ground-fault circuit-interrupter.

**(a) 8 (b) 10 (c) 15 (d) 20**

33. Portable appliances used on 15 or 20 amp branch circuits, the rating of any one portable appliance shall not exceed ____ percent of the branch circuit rating.

**(a) 60 (b) 100 (c) 80 (d) 50**

34. All fixtures installed in damp locations shall be marked ____.

**(a) waterproof (b) suitable for wet locations (c) damp locations (d) weatherproof**

35. What kind of lighting loads does the Code say there shall be no reduction in the size of the neutral conductor?

**(a) dwelling unit (b) hospital (c) nonlinear (d) motel**

36. How would you seal unused ko's in panels and boxes?

**(a) cardboard (b) duct seal (c) tape (d) metal plugs and plates**

37. Electrodes of steel or iron shall have a diameter of at least ____.

**(a) 1/2" (b) 3/4" (c) 1" (d) 5/8"**

38. Liquidtight flexible conduit shall not be permitted ____.

**(a) in hazardous locations**
**(b) in high temperature areas**
**(c) in exposed and concealed work**
**(d) where installations requires flexibility or protection from liquids, vapors or solids**

39. In closed construction in a manufactured building, cables shall be permitted to be secured only at cabinets, boxes, or fittings where ____ or smaller conductors are used and protected as required.

**(a) #2 AWG (b) #10 AWG (c) #2/0 AWG (d) #250 kcmil**

40. The maximum length of exposed cord in a fountain shall be ____ feet.

**(a) 3 (b) 4 (c) 6 (d) 10**

41. Fixture studs that are not part of outlet boxes, _____ shall be made of steel, malleable iron, or other material suitable for the application.

I. crowfeet   II. hickeys   III. tripods

**(a) I only   (b) II only   (c) III only   (d) I, II and III**

42. A garbage disposal in the kitchen of a residence provided with a type SO three-conductor cord terminated with a grounding-type attachment plug shall be permitted where all of the following conditions are met _____.

I. the receptacle shall be readily accessible
II. the receptacle shall be located to avoid physical damage to the flexible cord
III. the recptacle shall be accessible
IV. the length of the cord shall not be less than 18" and not over 36"

**(a) I, II and IV   (b) I, II and III   (c) II, III and IV   (d) III and IV**

43. The minimum radius of the inside of a bend for a 3/4" flexible metallic tubing used for flexing is _____ inches.

**(a) 17 1/2   (b) 12 1/2   (c) 10   (d) 5**

44. Adjacent load-carrying conductors have the dual effect of raising the _____ and impeding heat dissipation.

**(a) insulation rating   (b) heat above 86°F   (c) ambient temperature   (d) skin effect**

45. Cables of the AC type, except ACL, shall have an internal bonding strip of _____ in intimate contact with the armor for its entire length.

I. aluminum   II. copper

**(a) I only   (b) II only   (c) either I or II   (d) neither I nor II**

46. Which of the following statements about a #2 THHN cu conductor is correct?

**(a) its maximum operating temperature is 90° C**
**(b) it has a nylon insulation**
**(c) its area is .067 square inches**
**(d) it has a DC resistance of .319 ohms per m/ft.**

47. According to the Code, conductors on poles, where not placed on racks or brackets, shall be separated not less than ____ inches.

**(a) 6 (b) 12 (c) 18 (d) 24**

48. Fixtures shall be supported independently of the outlet box where the weight exceeds ____ pounds.

**(a) 60 (b) 50 (c) 40 (d) 30**

49. Every circuit breaker having an interrupting rating other than ____ amperes, shall have its interrupting rating shown on the breaker.

**(a) 1000 (b) 2000 (c) 5000 (d) 7500**

50. Hoistway is a ____ in which an elevator or dumbwaiter is designed to operate.

**(a) shaftway (b) hatchway (c) well hole (d) all of these**

# OPEN BOOK EXAM #9

# 50 QUESTIONS
# TIME LIMIT - 2 HOURS

**TIME SPENT** ☐ **MINUTES**

**SCORE** ☐ **%**

## JOURNEYMAN OPEN BOOK EXAM #9 Two Hour Time Limit

1. Where devices containing a disconnecting means are mounted out of reach, suitable means shall be provided to operate the disconnecting means from the floor. Which of the following is permitted?

**(a) devices cannot be mounted out of reach**
**(b) ladders**
**(c) sticks**
**(d) no method is permitted**

2. Tubing having cut threads and used as arms or stems on light fixtures may not be less than ____ inches wall thickness.

**(a) .040 (b) .050 (c) .010 (d) .005**

3. Ground-fault circuit-interrupters shall be installed in the branch circuit supplying underwater pool lighting fixtures operating at more than ____ volts.

**(a) 12 (b) 15 (c) 24 (d) 50**

4. Each transformer shall be provided with a nameplate giving the name of the manufacturer; rated kv; frequency; primary and secondary voltage; impedance of transformers ____ kva and larger.

**(a) 112 1/2 (b) 25 (c) 33 (d) 50**

5. ____ is defined as properly localizing a fault condition to restrict outages to the equipment affected, accomplished by choice of selective fault protective devices.

**(a) Monitoring (b) Coordination (c) Choice selection (d) Fault device**

6. Two-wire DC circuits and AC circuits of two or more ungrounded conductors shall be permitted to be tapped from the ungrounded conductors of circuits having ____.

**(a) a properly sized tap conductor**
**(b) less than 50 volts**
**(c) a balanced neutral system**
**(d) a grounded neutral conductor**

7. Application of demand factors to small appliance and laundry loads in dwellings are permitted in Table ____.

**(a) 220-3 (b) 220-11 (c) 220-13 (d) 220-20**



TH

8. Conductors for festoon lighting shall be of the ____ type.

I. thermoplastic  II. rubber covered  III. shielded

**(a) I only  (b) I or II only  (c) II or III only  (d) I, II, or III**

9. Not more than one conductor shall be connected to the grounding electrode by a single clamp or fitting unless the clamp or fitting is ____.

**(a) cast bronze or brass**
**(b) listed for multiple conductors**
**(c) 0.043" in thickness**
**(d) none of these**

10. FCC cable can have individual branch circuits with a rating not exceeding ____ amperes.

**(a) 15  (b) 20  (c) 25  (d) 30**

11. Auxiliary equipment for electric-discharge lamps shall be ____ and treated as sources of heat.

**(a) enclosed in noncombustible cases**
**(b) thermally protected**
**(c) weatherproof**
**(d) ventilated**

12. Where used outside, aluminum or copper-clad aluminum grounding conductors shall not be installed within ____ inches of earth.

**(a) 24  (b) 18  (c) 30  (d) 36**

13. A receptacle outlet installed outdoors shall be located so that ____ is not likely to touch the outlet cover or plate.

**(a) persons  (b) water accumulation  (c) metal  (d) none of these**

14. Time switches, flashers, and similar devices where mounted so they are accessible only to qualified persons and so located in an enclosure that any energized parts within ____ of the manual adjustment or switch are covered by suitable barriers.

**(a) 4"  (b) 6"  (c) 12"  (d) 18"**

15. What size rigid PVC conduit schedule 40 is required for eight #6 XHHW conductors?

**(a) 3/4" (b) 1" (c) 1 1/4" (d) 1 1/2"**

16. The minimum radius for a bend of 1" rigid conduit with three #10 TW conductors is ____ inches. (one shot bender)

**(a) 6 (b) 11 (c) 5 3/4 (d) none of these**

17. The feeder conductor ampacity shall not be lower than that of the service-entrance conductors where the feeder conductors carry the total load supplied by service-entrance conductors with an ampacity of ____ amperes or less.

**(a) 50 (b) 55 (c) 100 (d) 125**

18. Receptacles located ____ feet above the floor are not counted in the required number of receptacles along the wall.

**(a) 4 (b) 6 (c) 5 1/2 (d) none of these**

19. Pool-associated motors shall be connected to an equipment grounding conductor not smaller than # ____.

**(a) 14 (b) 12 (c) 10 (d) 8**

20. To qualify as a lighting and appliance branch circuit panelboard, the number of circuits rated 30 amperes or less with neutrals must be ____.

**(a) more than 10% (b) 42 or less (c) 24 or more (d) 10%**

21. What is the area of square inch for a #12 RHW without outer covering?

**(a) .0353 (b) .0293 (c) .182 (d) .026**

22. Metal enclosures used to protect ____ from physical damage shall not be required to be grounded.

**(a) service conductors (b) feeders (c) cable assemblies (d) none of these**

23. Connection devices or fittings must not connect grounding conductors to equipment by means of ____.

**(a) pressure connections**
**(b) solder**
**(c) lugs**
**(d) approved clamps**

24. A bare #4 conductor may be concrete encased and serve as the grounding electrode when at least ____ feet in length.

**(a) 10 (b) 12 (c) 20 (d) 15**

25. Which of the following is **not** a standard classification for a branch circuit supplying several loads?

**(a) 20 amp (b) 25 amp (c) 30 amp (d) 50 amp**

26. Underfloor raceways may be occupied up to ____ percent of the area.

**(a) 55 (b) 30 (c) 40 (d) 38**

27. The volume per #14 conductor required in a box is ____ cubic inch.

**(a) 2.25 (b) 2 (c) 3 (d) 2.5**

28. What size copper grounding electrode conductor is required for a #1500 kcmil copper service conductor?

**(a) #2/0 (b) #3/0 (c) #0 (d) #2**

29. Electrical nonmetallic tubing shall be clearly and durably marked at least every ____ feet.

**(a) 3 (b) 6 (c) 8 (d) 10**

30. Vertical and horizontal spacing between supported cablebus conductors shall not be less than ____ at the points of support.

**(a) 1" (b) 1 1/2" (c) 2" (d) one conductor diameter**

31. ____ switches shall be used for capacitor switching.

**(a) Isolation (b) Group-operated (c) Shunt (d) High-voltage**

32. Disconnecting means shal be accessible, located within sight from pool, and shall be located at least ____ horizontally from the inside walls of the pool.

**(a) 18" (b) 2' (c) 4' (d) 5'**

33. The secondary circuits of wound-rotor AC motors, including conductors, controllers, resistors, etc. shall be considered as protected against overload by the ____.

**(a) disconnect**
**(b) controller**
**(c) breaker**
**(d) motor-overload device**

34. Enclosures for overcurrent devices in damp or wet locations shall be identified for use in such locations and shall be mounted so there is at least ____ inch air space between the enclosure and the wall.

**(a) 1/4 (b) 3/8 (c) 3/4 (d) 1**

35. Which of the following is required for temporary wiring?

**(a) Flexible cords shall be protected from accidental damage.**
**(b) All branch circuits shall originate in an approved panelboard.**
**(c) All conductors shall be protected as provided in article 240.**
**(d) All of these.**

36. Nonmetallic surface extensions with one or more extensions shall be permitted to be run in any direction from an existing outlet, but not on the floor or within ____ inches from the floor.

**(a) 6 (b) 4 (c) 3 (d) 2**

37. Water heaters having a capacity of ____ gallons or less shall have a branch circuit rating not less than 125% of the rating of the water heater.

**(a) 60 (b) 75 (c) 90 (d) 120**

38. A spacing of not less than ____ shall be maintained between neon tubing and the nearest surface, other than its support.

**(a) 1/4" (b) 1/2" (c) 3/8" (d) 5/16"**

39. An autotransformer starter shall provide ____.

I. an "off position" II. a running position III. at least one starting position

**(a) I only (b) II only (c) I and II (d) I, II and III**

40. A metal elbow installed underground in a run of nonmetallic conduit is not required to be grounded, if it is isolated by a minimum over of at least ____ inches to any part of the elbow.

**(a) 6 (b) 12 (c) 18 (d) 24**

41. Temporary wiring shall be removed ____ upon completion of construction or purpose for which the wiring was installed.

**(a) 30 days (b) immediately (c) A.S.A.P. (d) 60 days**

42. Type MV cables shall not be used unless identified for the use ____.

I. in cable trays II. where exposed to direct sunlight

**(a) I only (b) II only (c) both I and II (d) neither I nor II**

43. In completed installations each outlet box shall have a ____.

**(a) receptacle (b) switch (c) cover (d) fixture**

44. Which of the following shall be provided where necessary to assure electrical continuity?

**(a) Grounding (b) Bonding (c) Jumpers (d) Shunts**

45. A continuous white or natural gray covering on a conductor shall be used only for the ____ conductor.

**(a) grounding (b) ungrounded (c) hot (d) grounded**

46. Grounding of a metal raceway used to protect Romex is required if the raceway is ____ feet or over, or within reach of ground or grounded metal.

**(a) 6 (b) 8 (c) 10 (d) 25**

47. A single electrode consisting of a ____ which does not have a resistance to ground of 25Ω or less shall be augmented by one additional electrode.

I. rod II. pipe III. plate

**(a) I only (b) II only (c) III only (d) I, II or III**

48. What is the va input of a fully loaded 5 hp 230 volt single-phase motor?

**(a) 746 (b) 3730 (c) 6440 (d) 12,880**

49. The minimum size of a copper equipment grounding conductor required for equipment connected to a 40 amp circuit is ____.

**(a) #12 (b) #14 (c) #8 (d) #10**

50. 2" rigid metal conduit shall be supported every ____ feet.

**(a) 10 (b) 12 (c) 14 (d) 16**

# OPEN BOOK EXAM #10

## 50 QUESTIONS
## TIME LIMIT - 2 HOURS

**TIME SPENT** ☐ **MINUTES**

**SCORE**  **%**

**JOURNEYMAN OPEN BOOK EXAM #10** **Two Hour Time Limit**

1. In dwelling units and guest rooms of hotels, motels, and similar occupancies, the voltage shall not exceed 120 volts, between conductors that supply the terminals of ____.

I. cord and plug connected loads 1440 volt amperes or less
II. cord and plug connected loads 1440 volt amperes or less, or less than 1/8 horsepower
III. lighting fixtures

**(a) I only (b) I and II only (c) I and III only (d) I, II and III**

2. Unless identified for use in the operating environment, no conductors or equipment shall be located in _____ having a deteriorating effect on the conductors or equipment.

I. damp or wet locations II. where exposed to gases, fumes, vapors, liquids, etc.

**(a) I only (b) II only (c) both I and II (d) neither I nor II**

3. Transformers of more than _____ kva rating shall be installed in a transformer room of fire-resistant construction.

**(a) 35,000 (b) 87 1/2 (c) 112 1/2 (d) 75**

4. Conduit bodies shall have a cross-sectional area at least _____ that of the largest conduit to which they are connected, #6 conductors and smaller.

**(a) 100% (b) twice (c) 40% (d) 75%**

5. Type FCC cable shall be clearly and durably marked on both sides at intervals of not more than _____.

**(a) 18" (b) 2' (c) 30" (d) 3'**

6. A system or circuit conductor that is intentionally grounded is a _____ conductor.

**(a) grounding (b) unidentified (c) grounded (d) none of these**

7. The area of square inches for a #1/0 bare conductor is _____.

**(a) .087 (b) .109 (c) .137 (d) .173**

8. _____ plugs driven into holes in masonry, concrete, plaster, or similar materials shall not be used.

**(a) Metal (b) Plastic (c) Leather (d) Wooden**

9. Thermal insulation shall not be installed within ____ inches of the recessed fixture enclosure.

**(a) 3 (b) 4 (c) 6 (d) 8**

10. Service entrance cables, where subject to physical damage, shall be protected in which of the following?

I. EMT II. IMC III. rigid metal conduit

**(a) III only (b) II and III (c) I, II and III (d) I and III**

11. Overhead conductors, not supported by messenger wires, for festoon lighting shall not be smaller than ____.

**(a) #14 (b) #12 (c) #10 (d) #8**

12. Which of the following is a standard size fuse?

**(a) 75 (b) 95 (c) 601 (d) 1500**

13. Where conductors are adjusted to compensate for voltage drop, equipment grounding conductors, where required, shall be adjusted proportionally according to ____.

**(a) diameter (b) cross section area (c) circular mil area (d) circumference**

14. Voltage between the hot (ungrounded) conductors on FCC cable shall not exceed ____ volts.

**(a) 50 (b) 300 (c) 150 (d) 600**

15. The work space required by the code for electrical equipment shall not be used for ___.

I. passageway II. storage III. panelboards

**(a) I only (b) II only (c) III only (d) I and II only**

16. Cablebus framework, where ____, shall be permitted as the equipment grounding conductor for branch circuits and feeders.

**(a) bonded as required by Article 250**
**(b) welded**
**(c) protected**
**(d) galvanized**

17. According to the Code, metal enclosures for grounding electrode conductors shall be ____.

**(a) not permitted (b) electrically continuous**
**(c) rigid conduit (d) none of these**

18. Feeders containing a common neutral shall be permitted to supply ____.

I. 2 or 3 sets of 3-wire feeders II. 2 sets of 4-wire or 5-wire feeders

**(a) I only (b) II only (c) either I or II (d) neither I nor II**

19. Operation at loads, and intervals of time, both of which may be subject to wide variation is the definition of ____.

**(a) varying duty (b) demand factor**
**(c) cycle (d) periodic duty**

20. Underground cable installed under a building shall be in a ____ that is extended beyond the outside walls of the building.

**(a) sleeve (b) duct bank (c) gutter (d) raceway**

21. Where NM cable is used, the cable assembly, including the sheath, shall extend into the box no less than ____.

**(a) 1/2" (b) 3/4" (c) 1/4" (d) 1"**

22. The current carried continuously in bare copper bars in auxiliary gutters shall not exceed ____ amperes per square inch.

**(a) 560 (b) 700 (c) 800 (d) 1000**

23. Under the optional method of calculation for a single-family dwelling, all "other load" beyond the initial 10 kva is to be assessed at ____ percent.

**(a) 40 (b) 50 (c) 60 (d) 75**

24. Metal conduit and metal piping within ____ feet of the inside walls of the pool and that are not separated from the pool by a permanent barrier are required to be bonded.

**(a) 4 (b) 5 (c) 8 (d) 10**

25. Suitable covers shall be installed on all boxes, fittings, and similar enclosures to prevent accidental contact with ____ parts or physical damage to parts or insulation. Over 600v nominal.

**(a) energized** **(b) mechanical**
**(c) electrical** **(d) none of these**

26. A unit of an electrical system which is intended to carry but not utilize electric energy would be a ____.

I. light bulb II. snap switch III. device IV. receptacle

**(a) I only (b) III only (c) I, II and IV (d) II, III, and IV**

27. Type ____ cable is a factory assembly of one or more conductors, each individually insulated and enclosed in a metallic sheath of interlocking tape, or a smooth or corrugated tube.

**(a) MI (b) AC (c) MC (d) MV**

28. ____ boxes shall not be used where conduits or connectors requiring the use of locknuts or bushings are to be connected to the side of the box.

**(a) Round (b) Shallow (c) Device (d) Gang**

29. Lampholders installed over highly combustible material shall be of the ____ type.

**(a) porcelain (b) low smoke (c) switched (d) unswitched**

30. Nonconductive coatings (such as paint, lacquer, and enamel) on equipment to be grounded shall be removed from threads and other contact surfaces to ____.

**(a) provide a water tight joint**
**(b) provide a sealed joint**
**(c) assure good electrical continuity**
**(d) lower inductance**

31. UF cable installed to an outdoor post light on a residential branch circuit rated 15 amps, 115 volt would require a minimum burial depth of ____ inches.

**(a) 24 (b) 18 (c) 12 (d) 6**

32. The ampacity of types NM and NMC cable shall be that of ____ conductors.

**(a) 60° C (b) 75° C (c) 90° C (d) 140° C**

33. When an outlet is removed from a cellular metal floor raceway, the sections of circuit conductors supplying the outlet shall be _____.

**(a) taped (b) dead-ended (c) shorted together (d) removed from the raceway**

34. Bored holes in wood members for cable or raceway-type wiring shall be bored so that the edge of the hole is not less than ____ from the nearest edge.

**(a) 1 1/4" (b) 1 1/8" (c) 1 1/2" (d) 1 1/16"**

35. Where practicable, dissimilar metals in contact anywhere in the system shall be avoided to eliminate the possibility of _____.

**(a) hysteresis (b) galvanic action (c) specific gravity (d) resistance**

36. The radius of the inner edge of any bend shall not be less than _____ times the diameter of the metallic sheath for cable not more than 3/4" in external diameter.

**(a) 5 (b) 3 (c) 8 (d) 10**

37. A 500 ampere load supplied by a 120/240v feeder requires a feeder neutral with an ampacity of _____ amps.

**(a) 410 (b) 340 (c) 280 (d) 350**

38. A service drop over a residential driveway shall have a minimum height of _____ feet.

**(a) 10 (b) 12 (c) 15 (d) 18**

39. The grounded conductors of _____ metal-sheathed cable shall be identified by distinctive marking at the terminals during the process of installation.

**(a) armored cable (b) mineral-insulated (c) copper (d) aluminum**

40. Electric heating appliances employing resistance-type heating elements rated more than _____ amperes shall have the heating elements subdivided.

**(a) 60 (b) 50 (c) 48 (d) 35**

41. What is the minimum size conductor permitted for general wiring under 600 volts?

**(a) #12 copper (b) #14 aluminum (c) #14 copper (d) #12 aluminum**

42. Class III locations are those that are hazardous because of ____.

**(a) the presence of combustible dust**
**(b) over 8' depth of water**
**(c) flammable gases or vapors may be present in the air**
**(d) the presence of easily ignitible fibers or flyings**

43. The maximum number of quarter bends in one run of EMT is ____.

**(a) two (b) four (c) five (d) none of these**

44. The conductors, including splices and taps in metal surface raceway shall not fill the raceway to more than ____ percent of its area at that point.

**(a) 75 (b) 40 (c) 38 (d) 53**

45. The minimum feeder load for a 40 foot long show window is ____ va.

**(a) 4000 (b) 8000 (c) 10,000 (d) none of these**

46. Type MC cable shall not be used where exposed to ____ conditions.

**(a) wet (b) destructive corrosive (c) unsafe (d) high-heat**

47. Where MI cable terminates, a ____ shall be provided immediately after stripping to prevent the entrance of moisture into the insulation.

**(a) bushing (b) connector (c) fitting (d) seal**

48. A nipple contains four #6 THW copper current-carrying conductors. The ampacity of each conductor would be ____ amperes.

**(a) 65 (b) 52 (c) 39 (d) 55**

49. The DC resistance @ 167° F for a #2/0 bare aluminum conductor would be ____ ohm per thousand feet of conductor.

**(a) 0.0967 (b) 0.101 (c) 0.319 (d) 0.159**

50. The approximate area of square inch for a #4/0 THW aluminum building wire is ____.

**(a) .3288 (b) .3904 (c) .3267 (d) .2780**

# OPEN BOOK EXAM #11

## 50 QUESTIONS
## TIME LIMIT - 2 HOURS

**TIME SPENT** ☐ **MINUTES**

**SCORE** ☐ **%**

**JOURNEYMAN OPEN BOOK EXAM #11** **Two Hour Time Limit**

1. Where a ____ supplies continuous loads or any combination of continuous and noncontinuous loads, the rating of the overcurrent device shall not be less than the noncontinuous load plus 125% of the continuous load.

**(a) load (b) branch-circuit (c) demand (d) conductor**

2. Type USE service entrance cable, identified for underground use in a cabled assembly, may have a ____ concentric conductor applied.

**(a) bare copper (b) covered métal**
**(c) bare aluminum (d) covered**

3. Throughout the Code, the voltage considered shall be that at which the circuit ____.

**(a) is grounded (b) feeds (c) operates (d) drops**

4. Conductors shall be considered outside a building ____.

I. when installed in a raceway
II. where installed within a building in a raceway enclosed by 2" of brick
III. where installed under not less than 2" of concrete beneath a building

**(a) II only (b) III only (c) II and III only (d) I, II and III**

5. The ampacity of capacitor circuit conductors shall not be less than ____ percent of the rated current of the capacitor.

**(a) 100 (b) 115 (c) 135 (d) 150**

6. The temperature rating of a conductor is the maximum temperature, at any location along its length, that the conductor can withstand over a prolonged time period without ____.

**(a) tripping the breaker (b) serious degradation**
**(c) short circuiting (d) a ground fault**

7. A grounding electrode conductor shall not be required for a system that supplies a ____ and is derived from a transformer not more than 1000 va.

**(a) Class I circuit (b) Class II circuit (c) Class III circuit (d) all of these**

8. Branch circuits in dwelling units shall supply only loads within that dwelling unit or loads associated only with that dwelling unit. Branch circuits required for the purpose of lighting, ____, or other needs for public or common areas shall not be supplied from a dwelling unit panelboard.

I. communications II. signal III. central alarm

**(a) I only (b) II only (c) III only (d) I, II and III**

9. A #16 fixture wire is considered protected by a 20 amp overcurrent device up to ____ feet.

**(a) 25 (b) 50 (c) 75 (d) 100**

10. Two or three single-pole switches or breakers, capable of individual operation, shall be permitted on multiwire circuits, one pole for each ungrounded conductor, as one multipole disconnect provided they are equipped with ____ to disconnect all conductors of the service with no more than six operations of the hand.

I. a master handle II. handle ties

**(a) I only (b) II only (c) both I and II (d) neither I nor II**

11. The ampacity of type UF cable shall be that of ____ conductors.

**(a) 60°F (b) 75°C (c) 140°C (d) 60°C**

12. Each fitting attached to a heavy-duty lighting track shall ____.

**(a) have individual overcurrent protection**
**(b) have double lock nuts**
**(c) be raintight**
**(d) not be over 3' in length**

13. What is the cross sectional area of a 1 1/2" rigid metal conduit?

**(a) 2.071 (b) .829 (c) 3.408 (d) 1.624**

14. Unless identified as suitable for use with infrared heating lamps, screw-shell lampholders shall not be used with infrared lamps over ____ watts rating.

**(a) 150 (b) 300 (c) 5000 (d) none of these**

15. What is the minimum thickness of metal for a 6" x 4" x 3 1/4" box?

**(a) .0625" (b) .0747" (c) 15 MSG (d) 16 MSG**

16. A receptacle which is secured solely by a single screw, installed in a raised cover on a four square box ____.

**(a) is prohibited in all cases**
**(b) is allowed without exception**
**(c) is allowed only for a receptacle listed for such use**
**(d) is allowed only when the raised cover is installed on a nonmetallic box**

17. A circuit containing #12 THHN conductors is a ____ rated circuit when protected by a 15 amp rated circuit breaker.

**(a) 25 amp (b) 20 amp (c) 15 amp (d) 30 amp**

18. A switch or circuit breaker should disconnect all grounded conductors of a circuit ____.

**(a) before it disconnects the ungrounded conductors**
**(b) after it disconnects the ungrounded conductors**
**(c) simultaneously as it disconnects the ungrounded conductors**
**(d) none of these**

19. Fixed appliances rated at not over ____ volt-amperes or 1/8 hp the branch-circuit overcurrent device shall be permitted to serve as the disconnecting means.

**(a) 240 (b) 300 (c) 400 (d) 480**

20. What is the ampacity of a #8 XHHW copper conductor in a wet location?

**(a) 55 amps (b) 50 amps (c) 45 amps (d) 40 amps**

21. Flexible metal conduit shall be secured by approved means at intervals not exceeding ____ feet and within 12" on each side of every outlet box.

**(a) 2 (b) 4 (c) 4 1/2 (d) 8**

22. A type of surface or flush raceway, designed to hold conductors and receptacles, is called ____.

**(a) underfloor raceway (b) cellular metal floor raceway**
**(c) multioutlet assembly (d) recessed outlets**

23. At what angle does a header attach to a floor duct?

**(a) reverse (b) parallel (c) right angle (d) none of these**

24. Loop wiring for underfloor raceways, shall not be considered ____.

**(a) a splice (b) a tap (c) both (a) and (b) (d) neither (a) nor (b)**

25. Induction heating coils that operate or may operate at a voltage greater than 30 volts AC shall be ____ to protect personnel in the area.

I. isolated
II. made inaccessible by location
III. enclosed in a nonmetallic enclosure
IV. enclosed in a split metallic enclosure

**(a) I or III only (b) I, II or III only (c) I, II or IV only (d) I, II, III or IV**

26. An office building has a 24 volt branch circuit installed for landscape lighting around the front of the building. The circuit was installed in UF cable which requires a minimum burial depth of ____ inches for this circuit.

**(a) 6 (b) 8 (c) 12 (d) 24**

27. Plaster, drywall or plasterboard surfaces that are broken or incomplete shall be repaired so there will be no gaps or open spaces greater than ____ inch at the edge of the fitting or box.

**(a) 1/16 (b) 1/8 (c) 3/16 (d) 1/4**

28. Concealed knob-and-tube wiring shall be permitted to be used only for extensions of existing installations and elsewhere only by special permission under the following conditions ____.

I. in unfinished attic and roof spaces when such spaces are insulated by loose or rolled insulating material
II. in the hollow spaces of walls and ceilings
III. in unfinished attic and roof spaces as provided in section 324-11

**(a) I only (b) I and II only (c) II and III only (d) I, II and III**

29. Raceways shall be installed ____ between outlet, junction or splicing points prior to the installation of conductors.

**(a) partially (b) complete (c) straight (d) tightly**

30. Flexible cords to portable electrically heated appliances rated at more than ____ watts shall be approved for heating cords.

**(a) 50 (b) 100 (c) 300 (d) 500**

31. A single grounding electrode is permitted when the resistance to ground does not exceed ____ ohms.

**(a) 5 (b) 10 (c) 15 (d) 25**

32. What is the area of square inches for a #12 RHH with an outer covering?

**(a) .212 (b) .0353 (c) .0437 (d) .0293**

33. Unfinished basements are defined as portions or areas of basements not intended as habitable rooms and ____.

I. work areas II. storage areas III. tool storage area

**(a) I only (b) II only (c) I and II only (d) I, II and III**

34. The ____, or other descriptive marking by which the organization responsible for the product may be identified, shall be placed on all electric equipment.

I. trademark II. cost III. manufacturer's name

**(a) I only (b) I and II only (c) I and III only (d) I, II and III**

35. The interior metal water piping system shall be bonded to the ____.

**(a) grounded conductor at the service**
**(b) grounding electrode conductor**
**(c) service equipment enclosure**
**(d) all of these**

36. Rigid schedule 80 PVC shall have a minimum burial depth of ____ inches.

**(a) 6 (b) 10 (c) 18 (d) 24**

37. Which of the following statements about FCC cable is **not** true?

**(a) a bottom shield shall be installed beneath all type FCC cable, connectors, and insulating ends**
**(b) FCC cable can cross over or under flat telephone cable**
**(c) an FCC system with a height above floor level exceeding 0.090 inches shall be tapered**
**(d) receptacles and connections need not be polarized**

38. Type AC cable shall be permitted for branch circuits and feeders in ____.

I. concealed work II. exposed work III. hazardous locations

**(a) I, II and III (b) II and III only (c) I and III only (d) I and II only**

39. Except by special permission, no conductor larger than ____ shall be installed in cellular metal floor raceways.

**(a) #1/0 (b) #2/0 (c) #250 kcmil (d) #500 kcmil**

40. Electrical continuity at service equipment shall be assured by ____.

I. threadless couplings and connectors made up tight for rigid metal conduit, IMC and EMT
II. threaded couplings and threaded bosses on enclosures with joints shall be made up wrenchtight where rigid metal conduit and IMC are involved
III. standard locknuts or bushings

**(a) I or III only (b) II or III only (c) I or II only (d) I, II or III**

41. The principal determinants of operating temperature are ____.

I. heat generated internally in the conductor as the result of load current flow
II. the rate at which generated heat dissipates into the ambient medium
III. adjacent load-carrying conductors
IV. ambient temperature

**(a) II and IV only (b) I and IV only (c) I, II and IV (d) I, II, III and IV**

42. The first floor of a building shall be that floor which is designed for human habitation and which has ____ percent or more of its perimeter level with or above finished grade of the exterior wall line.

**(a) 10 (b) 15 (c) 25 (d) 50**

43. Circuit breakers shall be so located or shielded so that persons ____.

**(a) will not be burned or otherwise injured by their operation**
**(b) other than the authority cannot locate them**
**(c) cannot operate them without a key**
**(d) other than the authority cannot remove them**

44. Electrical equipment such as a panelboard, shall include an exclusively dedicated space extending from the floor to a height of 6 feet or to the ____ whichever is lower. No piping, ducts, or equipment foreign to the electrical equipment shall be permitted in this dedicated space.

**(a) floor to suspended ceiling** **(b) structural ceiling**
**(c) wall to wall** **(d) basement to ceiling**

45. The ampacity of a device to open under short circuit or ground fault is based on its ____ rating.

**(a) operating (b) interrupting (c) ampacity (d) temperature**

46. Heavy-duty lampholders shall have a rating not less than ____ watts of the admedium type, and not less than ____ watts of any other type.

**(a) 750 ... 750 (b) 1000 ... 750 (c) 660 ... 750 (d) 660 ... 1000**

47. The minimum feeder-circuit conductor size, before the application of any adjustment or correction factors, shall have an allowable ampacity equal to or greater than the noncontinuous load plus ____ percent of the continuous load.

**(a) 100 (b) 125 (c) 80 (d) 75**

48. Conductive materials enclosing electrical conductors are grounded to ____.

I. prevent surges of voltage
II. prevent surges of lightning
III. to facilitate overcurrent device operation in case of ground faults

**(a) I only (b) II only (c) III only (d) all of these**

49. ____ cable shall be flame-retardant, moisture-resistant, fungus-resistant, and corrosion-resistant.

**(a) MI (b) USE (c) NMC (d) NM**

50. Circuit breakers shall not be located in the vicinity of easily ignitible material such as in ____.

**(a) hallways (b) laundry rooms (c) clothes closets (d) basements**

# OPEN BOOK EXAM #12

## 50 QUESTIONS
## TIME LIMIT - 2 HOURS

**TIME SPENT** ☐ **MINUTES**

**SCORE** ☐ **%**

**JOURNEYMAN OPEN BOOK EXAM #12** **Two Hour Time Limit**

1. What is the area of square inches for a #14 RHW without outer covering?

**(a) .0135 (b) .0209 (c) .0230 (d) .0327**

2. The ampacities provided by this section are based on temperature alone and do not take _____ into consideration.

**(a) insulation (b) AWG (c) CMA (d) voltage drop**

3. Flexible cords shall be secured to the undersides of showcases so that _____.

I. the free lead at the end of a group of showcases will have a female fitting not extending beyond the case
II. wiring will not be exposed to mechanical damage
III. a separation between cases not in excess of 2", nor more than 12" between the first case and the supply receptacle will be assured

**(a) I only (b) II only (c) III only (d) I, II and III**

4. All heating elements that are _____, and part of an appliance shall be legibly marked with the ratings in volts and amperes, or in volts and watts, or with the manufacturer's part number.

I. replaceable in the field II. rated over one ampere III. over 150 volts

**(a) II and III (b) I and II (c) I and III (d) I, II and III**

5. All conductors in a multiwire branch circuit shall originate from the same _____.

**(a) feeder (b) service (c) panelboard (d) receptacle**

6. The parallel conductors in each phase or neutral shall _____.

I. have the same insulation type and conductor material
II. be the same size in cma
III. be the same length and be terminated in the same manner

**(a) I only (b) II only (c) III only (d) I, II and III**

7. Where nails are used to mount knobs, they shall not be smaller than _____ penny.

**(a) 6 (b) 8 (c) 10 (d) 16**

8. In computing the load of fluorescent fixtures, the computation shall be based on the ____ of the fixture.

**(a) wattage of the ballast** **(b) wattage of the lamps**
**(c) total ampere rating** **(d) none of these**

9. Open conductors on insulators must be covered when they are within ____ feet of a building.

**(a) 10 (b) 12 (c) 15 (d) 25**

10. No grounded interior wiring shall be electrically connected to a supply system unless the supply system contains a corresponding conductor which is ____.

**(a) shielded (b) bonded (c) grounded (d) low-voltage**

11. All splices, joints and free ends of conductors are required to be covered with an insulation ____ the conductor.

**(a) as thick as (b) equivalent to (c) thicker than (d) larger than**

12. Appliances fastened in place, connected to branch circuits with other loads shall not exceed ____ percent of the branch circuit rating.

**(a) 40 (b) 50 (c) 70 (d) 80**

13. Multioutlet assembly may be used ____.

**(a) where concealed** **(b) in storage battery rooms**
**(c) in dry locations** **(d) in hoistways**

14. A multiwire branch-circuit may supply ____.

**(a) 120/240v to only one utilization equipment**
**(b) 120/240v where all ungrounded conductors are opened simultaneously**
**(c) both (a) and (b)**
**(d) neither (a) nor (b)**

15. Which of the following is **not** required on a motor nameplate?

**(a) watts (b) horsepower (c) manufacturer's identification (d) voltage**

16. Conductors shall **not** be installed in locations where the operating temperature will exceed that specified for the type of ____ used.

**(a) connectors (b) protection (c) insulation (d) wiring**

17. Where a service mast is used for the support of service drop conductors, it shall be of adequate strength or be supported by ____.

**(a) studs (b) braces or guys (c) rigid conduit (d) R.C. beams**

18. Permanent ladders or stairways shall be provided to give safe access to the working space around electric equipment over 600 volts installed on ____ or in attic or roof rooms or spaces.

I. balconies II. mezzanine floors III. platforms

**(a) I only (b) II only (c) III only (d) I, II and III**

19. The definition of a bathroom is an area including a ____ with one or more of the following: a toilet, a tub, or a shower.

**(a) water heater (b) sliding glass door (c) spa (d) basin**

20. Open conductors shall be supported on glass or porcelain-knobs, ____.

I. strain insulators II. brackets III. racks

**(a) I only (b) II only (c) III only (d) I, II or III**

21. Which of the following is **true** concerning type NM cable?

**(a) it may be installed where exposed to corrosive fumes**
**(b) it may be fished in air voids in masonry block or tile walls**
**(c) it may be embedded in masonry, concrete, or plaster**
**(d) it may be covered with plaster, adobe, or similar finish**

22. A three-wire, 240/120v single-phase 200 amp service for a dwelling requires what size THW copper conductors?

**(a) #4/0 (b) #3/0 (c) #2/0 (d) #1/0**

23. In dwelling units and guest rooms of hotels and motels, overcurrent devices shall not be located in ___.

**(a) hallways (b) bathrooms (c) bedrooms (d) kitchens**

24. Means shall be provided to disconnect the ____ of all fixed electric space heating equipment from all ungrounded conductors.

I. heater
II. motor
III. controller
IV. supplementary overcurrent protective devices

**(a) I and II only (b) II and IV only (c) I and IV only (d) I, III and IV only**

25. The grounded conductor (#1100 kcmil or less) brought to the service, shall ____ the minimum size grounding electrode conductor, sized from Table 250-94.

**(a) not be more than (b) not be less than (c) be twice (d) none of these**

26. Open motors with commutators shall be located so sparks cannot reach adjacent combustible material, but this ____.

**(a) is only required for over 600 volts**
**(b) shall not prohibit these motors on wooden floors**
**(c) does not prohibit these motors from Class I locations**
**(d) none of these**

27. The maximum number of overcurrent devices that may be installed in a lighting panel is ____.

**(a) 24 (b) 36 (c) 42 (d) 48**

28. Where an AC system operating at less than ____ volts is grounded at any point, the grounded conductor shall be run to each service.

**(a) 300 (b) 600 (c) 1000 (d) 1500**

29. General-use snap switch suitable only for use on alternating-current circuits for controlling ____.

I. resistive and inductive loads not exceeding the ampere rating of the switch
II. tungsten-filament lamp loads not exceeding the ampere rating of the switch
III. motor loads not exceeding 80% of the ampere rating of the switch

**(a) I only (b) III only (c) I and II only (d) I, II and III**

30. Type FCC cable wiring system is designed for installations under ____.

**(a) tile (b) carpet (c) carpet squares (d) concrete**

31. A residence has a front entrance on the north side of the house along with an attached garage with an 8' wide door, also a back entrance door on the south side of the house. How many lighting outlets are required for these outdoor entrances?

**(a) 1 (b) 2 (c) 3 (d) none of these**

32. Where required, drawings for feeder installations must be submitted before ____.

**(a) completion of installation** **(b) beginning of installation**
**(c) the use of feeders** **(d) the use of branch-circuits**

33. Nonmetallic sheath cable: If the attic is **not** accessible by stairs or permanent ladder, the cable needs to be protected only within ____ feet of a scuttle hole.

**(a) 2 (b) 3 (c) 6 (d) 10**

34. Transformer enclosures which extend directly to underwater pool light forming shells shall be provided with ____ grounding terminals.

**(a) one**
**(b) two**
**(c) the number of conduit entries plus one**
**(d) a grounding bus for**

35. Conductors other than service conductors shall not be installed in the same service raceway or service-entrance cable except ____.

I. grounding conductors II. load management control conductors having overcurrent protection

**(a) I only (b) II only (c) both I and II (d) neither I nor II**

36. A #6 copper conductor with one end bonded to the service raceway or equipment and with ____ inches or more of the other end made accessible on the outside wall of the dwelling is an example of the approved means for the external connection of a bonding, or grounding conductor to the service raceway or equipment.

**(a) 6 (b) 12 (c) 24 (d) 36**

37. Surge arresters shall be permitted to be located ____ and shall be made inaccessible to unqualified persons unless listed for installation in accessible location.

I. outdoors II. indoors

**(a) I only (b) II only (c) either I or II (d) neither I nor II**

38. Which of the following is **not** true concerning temporary wiring?

**(a) all lamps shall be protected by a suitable fixture or guard**
**(b) handle ties are permitted to disconnect multiwire branch circuits**
**(c) tests shall be performed on cords and receptacles and plugs for correct attachment to the equipment grounding conductor**
**(d) temporary power for Christmas decorative lighting shall not exceed 60 days**

39. Where single conductors or multiconductor cables are stacked or bundled longer than ____ without maintaining spacing and are not installed in raceways, the ampacity of each conductor shall be reduced.

**(a) 12" (b) 18" (c) 20" (d) 2'**

40. A previously unwired portion of an existing residence or a structural addition to an existing residence, either which exceeds ____ square feet, shall be computed in accordance with section 220-3c.

**(a) 100 (b) 250 (c) 300 (d) 500**

41. Where buildings exceed 3 stories or 50 feet in height, overhead lines shall be arranged, where practicable, so that a clear space (or zone) of at least ____ feet wide will be left either adjacent to the buildings or beginning not over 8 feet from them to facilitate the raising of ladders when necessary for fire fighting.

**(a) 4 (b) 6 (c) 8 (d) 10**

42. For uniform application of Articles 210, 215 and 220, a nominal voltage of ____ shall be used in computing the ampere load on a conductor.

**(a) 110/220 (b) 115/230 (c) 120/240 (d) 125/250**

43. When balancing a 3-wire circuit, single-phase 230/115 volt, the neutral conductor ____.

**(a) is used only for grounding**
**(b) should carry the unbalance**
**(c) should carry the sum**
**(d) none of these**

44. A motor control circuit ____.

I. carries electric signals to the controller, and carries the main power
II. does not carry electric signals to the controller, but carries the main power
III. carries the electric signals to the controller, but does not carry main power

**(a) I only (b) II only (c) III only (d) none of these**

45. When an outlet from an underfloor raceway is discontinued, the circuit conductors supplying the outlet _____.

**(a) may be handled like abandoned outlets on loop wiring**
**(b) may be reinsulated**
**(c) may be spliced**
**(d) shall be removed from the raceway**

46. If festoon lighting exceeds _____ feet, the conductors shall be supported by messenger wire.

**(a) 15 (b) 20 (c) 25 (d) 40**

47. A single receptacle shall have a rating of _____ percent of the branch-circuit rating.

**(a) 70 (b) 80 (c) 100 (d) 125**

48. An auxiliary gutter shall not extend a greater distance than _____ feet.

**(a) 10 (b) 30 (c) 50 (d) 75**

49. Solid dielectric insulated conductors operated above 2000 volts in permanent installations shall have ozone-resistant insulation and shall be _____.

**(a) covered (b) protected (c) shielded (d) surface mounted**

50. Conductor sizes are given in AWG and _____.

**(a) length (b) numbers (c) CM (d) insulation**

# OPEN BOOK EXAM #13

## 50 QUESTIONS
## TIME LIMIT - 2 HOURS

**TIME SPENT** ☐ **MINUTES**

**SCORE** ☐ **%**

## JOURNEYMAN OPEN BOOK EXAM #13 Two Hour Time Limit

1. For voltage of 600 or less, individual open service conductors in dry locations should be separated by _____.

**(a) 4 1/2" (b) 3 1/2" (c) 4" (d) 2 1/2"**

2. Where motors are provided with thermal housing, the housing shall be _____ and of substantial construction.

I. plastic II. metal

**(a) I only (b) II only (c) both I and II (d) neither I nor II**

3. A 1000 watt incandescent lamp shall have a _____ base.

**(a) mogul (b) standard (c) admedium (d) copper**

4. The internal depth of outlet boxes intended to enclose flush devices shall be at least _____.

**(a) 1/2" (b) 7/8" (c) 15/16" (d) 1 1/2"**

5. Ground-fault circuit protection for personnel is required for all 120v single-phase, 15 and 20 ampere receptacles that are installed in a dwelling unit _____.

**(a) attic (b) garage (c) bedroom (d) living room**

6. For dwelling unit(s), the computed floor area shall not include _____.

I. carports II. garages III. bathrooms IV. open porches

**(a) II and IV only (b) I, III and IV only (c) I, II and IV only (d) I, II, III and IV**

7. Which of the following is **not** true concerning the optional method for a dwelling unit?

**(a) The optional method of calculation is permitted if the service-entrance conductors have an ampacity of 200 or greater.**
**(b) The neutral would be determined by section 220-22.**
**(c) A demand of 40% of the nameplate rating(s) of electric space heating of four or more separately controlled units can be applied.**
**(d) A demand of 65% of the nameplate rating(s) of central electric space heating can be applied.**

8. Bonding all piping and ____ within the premises will provide additional safety.

**(a) water heaters (b) pumps (c) metal air ducts (d) none of these**

9. Finished ceilings containing heating cables shall be permitted to be covered with ____.

I. wallpaper II. plastic III. paint IV. wood

**(a) I or III (b) III or IV (c) I, III or IV (d) I, II, III or IV**

10. Metal-enclosed busways shall be installed so that ____ from induced circulating currents in any adjacent metallic parts will not be hazardous to personnel or constitute a fire hazard.

**(a) stray currents (b) magnetic flux (c) the impedance (d) temperature rise**

11. The largest conductor permitted in 3/8" flexible conduit is ____.

**(a) #12 (b) #16 (c) #14 (d) #10**

12. AC - DC general use snap switches may be used for control of inductive loads not exceeding ____ of the rating at the voltage.

**(a) 50% (b) 80% (c) 100% (d) 70%**

13. No point along the floor line in any useable wall space in a dwelling may be more than ____ feet from an outlet.

**(a) 6 (b) 6 1/2 (c) 8 (d) 10**

14. ____ conductors shall be used for wiring on fixture chains and other moveable parts.

**(a) Solid (b) Covered (c) Insulated (d) Stranded**

15. Overhead service drop conductors shall have a horizontal clearance of ____ feet from a pool.

**(a) 8 (b) 10 (c) 15 (d) 20**

16. The Code rules and provisions are enforced by ____.

**(a) the electric utility company that provides the power**
**(b) the U.S. government**
**(c) government bodies exercising legal jurisdiction over electrical installations**
**(d) U.L.**

17. Where permissible, the demand factor applied to that portion of the unbalanced neutral feeder load in excess of 200 amps is _____ percent.

**(a) 40 (b) 80 (c) 70 (d) 125**

18. Panelboards equipped with snap switches rated at 30 amps or less, shall have overcurrent protection not in excess of _____ amps.

**(a) 150 (b) 300 (c) 100 (d) 200**

19. Non-heating leads of heating cables operating in 208v systems, shall have a _____ color.

**(a) red (b) blue (c) yellow (d) brown**

20. Parts that must be removed for lamp replacement shall be _____.

I. insulated II. hinged III. held captive

**(a) I only (b) II or III (c) II only (d) I, II or III**

21. Flexible cord shall be permitted _____.

I. to facilitate the removal or disconnection of appliances
II. for connection of appliances to prevent the transmission of noise

**(a) I only (b) II only (c) both I or II (d) neither I nor II**

22. Messenger wires used to support festoon wiring shall **not** be attached to any _____.

I. plumbing equipment II. downspout III. fire escape

**(a) I only (b) II only (c) III only (d) I, II and III**

23. Flexible cords shall be connected to devices and to fittings so that tension will not be transmitted to joints or terminal screws. This shall be accomplished by _____.

**(a) special fitting designed for this**
**(b) winding with tape**
**(c) knot in cord**
**(d) all of these**

24. Service heads for service conductors shall be ____.

**(a) raintight (b) weatherproof (c) rainproof (d) watertight**

25. Open conductors run individually as service drops shall be ____.

I. insulated II. bare III. covered

**(a) I only (b) II only (c) III only (d) I or III**

26. What length of nipple may utilize the 60% conductor fill?

**(a) 12" (b) 18" (c) 24" (d) all of these**

27. A one-family dwelling unit that is at grade level shall have ____ outdoors.

**(a) one receptacle at the back**
**(b) one receptacle at the front**
**(c) two receptacles at the back**
**(d) one receptacle at front and one at the back**

28. The largest standard cartridge fuse rating is ____ amps.

**(a) 6000 (b) 1200 (c) 1000 (d) 600**

29. Surface metal raceways when extended through walls or floors must be in ____ lengths.

**(a) 8 foot (b) 3 foot (c) 5 foot (d) none of these**

30. Conductors shall be ____ unless otherwise provided in the Code.

**(a) lead (b) stranded (c) copper (d) aluminum**

31. What is the minimum size fixture wire?

**(a) #16 (b) #18 (c) #20 (d) #22**

32. The number of square feet that each plate electrode should present to the soil is ____ sq.ft.

**(a) 4 (b) 3 (c) 2 (d) 1**

33. Lighting systems operating at 30 volts or less shall be supplied from a maximum ___ ampere branch circuit.

**(a) 15 (b) 20 (c) 25 (d) 30**

34. The path to ground from circuits, equipment, and conductor enclosures shall ____.

I. have sufficiently low impedance to limit the voltage to ground and to faciliate the operation of the circuit protective devices in the circuit
II. shall be capable of safely carrying the maximum fault current likely to be imposed on it
III. be permanent and electrically continuous

**(a) I only (b) II only (c) III only (d) I, II and III**

35. Receptacles on construction sites shall not be installed on branch circuits which ____.

**(a) are over 15 amps (b) supply temporary lighting**
**(c) are supplied with cords (d) none of these**

36. Where screws are used to mount knobs, or where nails or screws are used to mount cleats, they shall be of a length sufficient to penetrate the wood to a depth to at least ____ the height of the knob and the full thickness of the cleat.

**(a) twice (b) one-half (c) one-quarter (d) 3 times**

37. FCC cable shall consist of ____ flat copper conductors, one of which shall be an equipment grounding conductor.

I. three II. four III. five

**(a) I only (b) II only (c) III only (d) I, II or III**

38. Where a metallic underfloor raceway system provides for the termination of an equipment grounding conductor, ____ shall be permitted.

I. EMT II. rigid PVC III. electrical nonmetallic tubing

**(a) I only (b) I or II (c) II or III (d) I, II or III**

39. When counting the number of conductors in a box, a conductor running through the box is counted as ____ conductor(s).

**(a) one (b) two (c) zero (d) none of these**

40. You may install ____ #8 TW conductors in a 1 1/2" E.M.T. conduit.

**(a) 13 (b) 22 (c) 18 (d) none of these**

41. Service cables mounted in contact with a building shall be supported at intervals not exceeding ____ feet.

**(a) 10 (b) 3 (c) 2 1/2 (d) 4 1/2**

42. Expansion joints and telescoping sections of raceways shall be made electrically continuous by equipment ____ or other means approved for the purpose.

**(a) grounding conductors (b) grounded conductor**
**(c) bonding jumpers (d) none of these**

43. Conductors ____ and larger shall be stranded when installed in raceways.

**(a) #10 (b) #8 (c) #6 (d) #4**

44. For the kitchen small appliance branch circuit in a dwelling, the Code requires not less than which of the following?

**(a) two 20 amp circuits (b) one 15 amp circuit**
**(c) two 15 amp circuits (d) one 20 amp circuit**

45. In combustible walls or ceilings, the front edge of an outlet box or fitting may set back of the finished surface ____.

**(a) 1/4" (b) 1/8" (c) 1/2" (d) not at all**

46. Lighting fixtures mounted on walls shall be installed with the top of the fixture lens at least ____ below the normal water level of the pool.

**(a) 15" (b) 3' (c) 18" (c) 12"**

47. Which of the following may not be used in damp or wet locations?

**(a) type AC armored cable (b) open wiring**
**(c) electrical metal tubing (d) rigid metal conduit**

48. A grounding electrode conductor subject to severe physical damage shall be protected when:

I. #4 or larger II. #6 or larger

**(a) I only (b) II only (c) both I and II (d) neither I nor II**

49. Which of the following is **not** a standard size fuse?

**(a) 110 amp (b) 601 amp (c) 75 amp (d) 125 amp**

50. A listed motor-circuit switch rated in horsepower for Design E motors rated greater than 2 hp, the motor circuit switch shall ____.

I. be not less than 1.3 times the rating of the motor rated over 100 hp
II. have a hp rating not less than 1.4 times the rating of a motor rated 3 - 100 hp
III. be marked as rated for use with Design E motors

**(a) I only (b) II only (c) III only (d) I, II, and III**

# OPEN BOOK EXAM #14

## 50 QUESTIONS
## TIME LIMIT - 2 HOURS

**TIME SPENT** ☐ **MINUTES**

**SCORE** ☐ **%**

**JOURNEYMAN OPEN BOOK EXAM #14** **Two Hour Time Limit**

1. Where a change occurs in the size of the ungrounded conductor, a similar change may be made in the size of the ____ conductor.

**(a) hot (b) grounding (c) grounded (d) none of these**

2. The current carried continuously in bare aluminum bars in auxiliary gutters shall not exceed ____ amperes per square inch.

**(a) 560 (b) 700 (c) 800 (d) 1000**

3. Type UF cable shall be permitted for ____.

**(a) service entrance cable (b) embedded in concrete**
**(c) direct burial (d) hoistways**

4. Soldered splices must be ____ so as to be electrically secure before soldering.

**(a) tinned (b) joined mechanically (c) taped (d) insulated**

5. No conductor larger than ____ shall be installed in a cellular concrete floor raceway without special permission.

**(a) #2 (b) #4 (c) #1/0 (d) #1**

6. In general, the voltage limitation between conductors in surface metal raceways is ____ volts.

**(a) 300 (b) 500 (c) 600 (d) 1000**

7. A nipple contains 6 - #6 THW copper current-carrying conductors. The ampacity of each conductor would be ____ amperes.

**(a) 65 (b) 50 (c) 52 (d) 45.5**

8. Conduit used to protect direct buried cable shall be provided with a ____ where the cable leaves the conduit underground.

**(a) seal (b) clamp (c) bushing (d) connector**

9. Temporary electrical power and lighting installations shall be permitted ____.

I. for developmental work
II. for permanent wiring
III. during emergencies and for tests

**(a) I only (b) II only (b) I and II only (d) I and III only**

10. Metal components of the FCC system shall be ____.

I. insulated from contact with corrosive substances
II. coated with corrosion-resistant materials
III. corrosion-resistant

**(a) I only (b) II only (c) III only (d) I, II or III**

11. An autotransformer which is used to raise the voltage to more than ____ volts, as part of a ballast for supplying lighting units, shall be supplied only by a grounded system.

**(a) 300 (b) 150 (c) 125 (d) 50**

12. ____ is a system in which heat is generated on the inner surface of a ferromagnetic envelope embedded in or fastened to the surface to be heated.

**(a) Duct heaters (b) Electrode-type boilers (c) Space heating (d) Skin effect heating**

13. The service disconnecting means shall plainly indicate ____.

**(a) its voltage rating (b) the maximum horsepower rating**
**(c) the maximum fuse size (d) whether it is in the open or closed position**

14. Using the optional method of calculation for a single-dwelling unit, the central space heating would be calculated at ____ percent.

**(a) 40 (b) 50 (c) 65 (d) 100**

15. Using the general method of calculation what is the minimum demand for a household clothes dryer?

**(a) 4 kw (b) 4.5 kw (c) 5 kw (d) 6 kw**

16. Type THW insulation has a ____ degree C rating for use in wiring through fixtures.

**(a) 60 (b) 75 (c) 85 (d) 90**

17. Flexible cord shall be considered as protected by a 20 amp branch circuit breaker if it is ____.

**(a) not less than 6' in length** **(b) #20 or larger**
**(c) #18 or larger** **(d) #16 or larger**

18. Service bonding jumpers must be sized ____.

**(a) according to the fuse size** **(b) same as the largest service conductor**
**(c) 1/3 as large as the service conductor** **(d) according to Table 250-66**

19. Unless specified otherwise, live parts of electrical equipment operating at ____ volts or more shall be guarded.

**(a) 32 (b) 50 (c) 115 (d) 150**

20. The frame of an electric range may be grounded by being connected to the grounded conductor of the 120/240v branch circuit, if the grounded conductor is not less than a ____ copper.

**(a) #10 (b) #8 (c) #6 (d) none of these**

21. The Code ____.

**(a) is not intended for a design specification**
**(b) is not intended for an instruction manual for untrained persons**
**(c) does not include installations in powerhouses under the exclusive control of electric utilities**
**(d) all of the above**

22. Bathroom receptacle outlets shall be supplied by ____ .

I. ground fault protection for personnel II. at least one 20 amp branch circuit

**(a) I only (b) II only (c) both I and II (d) neither I nor II**

23. The circular mil area of a #12 conductor is ____.

**(a) 10380 (b) 26240 (c) 6530 (d) 6350**

24. A lighting and appliance panelboard contains six 3-pole circuit breakers and eight 2-pole circuit breakers. The maximum allowable number of single-pole breakers permitted to be added in this panelboard is ____.

**(a) 8 (b) 16 (c) 28 (d) 42**

25. A 50 hp 208v, three-phase squirrel cage motor has a full-load current of ____ amps.

**(a) 130 (b) 143 (c) 162 (d) 195**

26. Where conductors of different systems are installed in the same raceway, one system shall have a neutral having an outer covering of white or natural gray and each other system having a neutral shall have an outer covering of ____.

**(a) white with green stripe**
**(b) white or natural gray**
**(c) blue**
**(d) white with colored stripe (other than green) or distinguished by other suitable means**

27. A feeder tap in a raceway terminating in a single circuit breaker with an ampacity 1/3 of the feeder conductors may extend not over ____ feet.

**(a) 6 (b) 10 (c) 25 (d) 50**

28. For general motor application the motor branch circuit fuse size must be determined from ____.

**(a) motor nameplate current (b) NEMA standards**
**(c) NEC Tables (d) Factory Mutual**

29. Minimum and maximum sizes of EMT are ____ except for special installations.

**(a) 5/16" to 3" (b) 3/8" to 4" (c) 1/2" to 3" (d) 1/2" to 4"**

30. Locations of lamps for outdoor lighting shall be ____.

I. below all energized conductors II. below all transformers

**(a) I only (b) II only (c) both I and II (d) neither I nor II**

31. The number and size of conductors in any raceway shall not be more than will permit ____.

I. ready installation or withdrawal of the conductors without damage to the conductors or to their insulation
II. dissipation of the heat

**(a) I only (b) II only (c) both I and II (d) neither I nor II**

32. Type MV cables shall be permitted for use on power systems rated up to ____ volts.

**(a) 600 (b) 4160 (c) 2300 (d) 35,000**

33. Handles or levers of circuit breakers, and similar parts which may move suddenly in such a way that persons in the vicinity are likely to be injured by being struck by them, shall be ____.

I. concealed II. isolated III. guarded

**(a) I or II only (b) II or III only (c) I or III only (d) I, II or III**

34. In a dwelling, the minimum feeder neutral for a 5 kva clothes washer/dryer would be ____ kva.

**(a) 5 (b) 4.3 (c) 3.5 (d) 3.0**

35. The grounding electrode shall be installed such that ____ of length is in contact with the soil.

**(a) 6' (b) 7' (c) 7' 6" (d) 8'**

36. Differences in inductive reactance and unequal division of current can be minimized by ____.

I. orientation of conductors
II. methods of construction
III. choice of materials

**(a) I only (b) II only (c) III only (d) I, II and III**

37. Connection from any grounding conductor of the type FCC cable shall be made to the shield system at each ____.

**(a) receptacle (b) outlet (c) switch (d) junction**

38. Where a metal lampholder is attached to a flexible cord, the inlet shall be equipped with an insulating bushing which, if threaded, shall not be smaller than nominal ____ inch pipe size.

**(a) 1/4 (b) 3/8 (c) 1/2 (d) 5/8**

39. The connection of a grounding electrode conductor to a driven ground rod shall be ____.

**(a) visible (b) accessible (c) readily accessible (d) not required to be accessible**

40. A thermal protector is intended to protect a motor against ____.

**(a) dangerous overheating** **(b) short circuit**
**(c) ground fault** **(d) none of these**

41. A 3" x 2" x 2" device box is how many cubic inches?

**(a) 12 (b) 14 (c) 10 (d) 8**

42. The power supply cord to a mobile home must not be longer than ____ feet.

**(a) 21 (b) 26 1/2 (c) 36 1/2 (d) 50**

43. Which of the following statements about the protection of nonmetallic sheathed cable from physical damage is/are correct?

I. when passing through a floor the cable shall be enclosed in a pipe or conduit extending at least 6 inches above the floor
II. when run across the top of the floor joists in an accessible attic, the cable shall be protected by guard strips

**(a) I only (b) II only (c) both I and II (d) neither I nor II**

44. The minimum clearance for service drops, not exceeding 600 volts, over commercial areas subject to truck traffic is ____ feet.

**(a) 10 (b) 12 (c) 15 (d) 18**

45. Plug fuses of the Edison-base type shall be used ____.

**(a) where overfusing is necessary**
**(b) only for 50 amps and above**
**(c) as a replacement for type S fuses**
**(d) only as a replacement item in existing installations**

46. In each kitchen and dining area a receptacle outlet shall be installed at each counter space ____ inches or wider.

**(a) 12 (b) 24 (c) 36 (d) 48**

47. Straight runs of 1 1/4" rigid metal conduit may be secured at not more than ____ intervals.

**(a) 5' (b) 10' (c) 12' (d) 14'**

48. The Code has assigned the color ____ to the high-leg of a 4-wire delta connected secondary.

**(a) black (b) red with green tracer (c) orange (d) pango pink**

49. When determining the load on the "volt-amps per square foot" basis, the floor area shall be computed from the ____ dimensions of the building.

**(a) inside (b) outside (c) midpoint (d) any of these**

50. In areas where the walls are frequently washed, conduit should be mounted with a ____ air space between the wall and the conduit.

**(a) 1/8" (b) 1/4" (c) 3/8" (d) 1/2"**

# ANSWERS

## ANSWERS JOURNEYMAN CLOSED BOOK EXAM #1

1. **(b)** II only
2. **(b)** capacitance
3. **(d)** equipment DEF 100
4. **(b)** general use DEF 100
5. **(d)** bonding DEF 100
6. **(b)** micrometer
7. **(d)** I,II,III, or IV
8. **(d)** I,II or III
9. **(b)** 5 full threads 501-4a1
10. **(a)** inverse time DEF 100
11. **(c)** I & II only 90-7
12. **(b)** separately derived DEF 100
13. **(c)** II & III only 110-3b
14. **(b)** overcurrent devices 210-3
15. **(d)** I,II & III 240-8 430-51
16. **(a)** unity 1.0
17. **(b)** II only ENT
18. **(d)** I & II only 400-14
19. **(d)** 90% efficiency for transformer
20. **(b)** skin effect
21. **(a)** nominal voltage DEF 100
22. **(d)** synchronous
23. **(b)** contact resistance
24. **(d)** I & III only DEF 100
25. **(c)** PVC schedule 40 300-5d
26. **(a)** insufficient resistive loads
27. **(c)** six steps 210-70a
28. **(c)** III only sinusoidal voltage
29. **(d)** greater is false
30. **(d)** all of these
31. **(d)** 40°C
32. **(c)** universal motor
33. **(a)** white or gray
34. **(b)** 100% 210-19a
35. **(a)** isolating switch DEF 100
36. **(d)** I,II, or III 300-8
37. **(c)** isolated DEF 100
38. **(c)** burnished
39. **(b)** power factor meter
40. **(a)** voltage & current
41. **(b)** 3 amps 200/5 = 40 ratio 120/40 = 3a
42. **(a)** shorted
43. **(a)** AC current flows
44. **(c)** 17/24 hp 1/3 + 1/4 + 1/8
45. **(c)** not through holes 400-8
46. **(c)** 50 pounds 410-16a
47. **(d)** emf electromotive force
48. **(a)** reduced voltage drop
49. **(d)** 900 ohms
50. **(d)** 160 turns 120/480=1/4 ratio = 40/160

## ANSWERS JOURNEYMAN CLOSED BOOK EXAM #2

1. **(a)** hertz
2. **(d)** resistance
3. **(b)** Z is impedance
4. **(c)** inductive load
5. **(c)** the splicing is easier
6. **(c)** impregnated paper
7. **(c)** cutting the lines of force
8. **(d)** field current
9. **(c)** separately excited
10. **(d)** reverse F1 & F2
11. **(c)** II and IV only
12. **(d)** ground rod 250-52c
13. **(d)** foot candles
14. **(c)** explosion proof
15. **(b)** volt amps
16. **(c)** 2238 watts 746w x 3 hp
17. **(d)** chemical reaction
18. **(b)** II & III only
19. **(d)** all of these
20. **(b)** skin effect
21. **(a)** to assure equipment grd. 300-10
22. **(c)** infinity
23. **(a)** series
24. **(d)** carry continuously
25. **(a)** screw shell 410-23
26. **(c)** 3 electrical rotations
27. **(b)** same as volt per turn
28. **(d)** not a Code requirement 250-56
29. **(c)** is suitable for charging batteries
30. **(a)** between white & black wire
31. **(c)** 120° separate each phase
32. **(d)** varying duty DEF 100
33. **(c)** surrounding the conductor DEF 100
34. **(a)** reactive power is decreased
35. **(b)** prevent chemical reactions
36. **(a)** excess of electrons
37. **(d)** result in damage to the ballast
38. **(a)** operation independent
39. **(d)** efficiency = output divided by input
40. **(b)** E x I x Time
41. **(d)** friction
42. **(a)** lines cut per second
43. **(a)** reactive power is decreased
44. **(c)** peak
45. **(c)** AC can be changed with transformer
46. **(b)** high starting torque
47. **(b)** induction
48. **(c)** change in voltage
49. **(b)** voltage applied
50. **(a)** greater the current flow

## ANSWERS JOURNEYMAN CLOSED BOOK EXAM #3

1. **(d)** I, II and III
2. **(c)** good PF **not** true
3. **(c)** ohms
4. **(c)** 100a 230-79c
5. **(b)** parallel
6. **(b)** green as hot, **not** true
7. **(d)** written consent DEF 100
8. **(c)** 1 megavolt
9. **(c)** both
10. **(c)** **not** true 210-3
11. **(c)** whenever current flows in conductor
12. **(a)** commutator
13. **(c)** 7.5 25 x 60w =1500 x 5 = 7500/1000
14. **(d)** machine
15. **(d)** I, II & III
16. **(c)** neutral carries the unbalance
17. **(b)** counterclockwise
18. **(a)** turn on another circuit
19. **(c)** current lag voltage, **not** true
20. **(d)** 1.0 unity
21. **(b)** variable
22. **(a)** layers of iron sheets
23. **(d)** limit excess voltage
24. **(c)** rate of work performed
25. **(b)** 70.7%
26. **(c)** I & III only
27. **(c)** PVC 24", **not** true T. 300-5
28. **(a)** equal currents in parallel
29. **(b)** lagging of magnetism
30. **(a)** voltage
31. **(b)** measure of ease of magnetism
32. **(c)** resistance
33. **(c)** either I or II
34. **(b)** reduce to simplest form
35. **(c)** causing AC to be generated
36. **(d)** 410-15a
37. **(d)** toggle bolt
38. **(c)** 1/4 as much
39. **(d)** I, II & III
40. **(b)** keep the surface clean
41. **(b)** static electricity
42. **(a)** 1" of concrete
43. **(c)** both
44. **(b)** special tools to make the joint
45. **(d)** I, II & III
46. **(b)** **not** true, 210-9 ex. 1, 2
47. **(a)** 25% 430-24a
48. **(d)** Y
49. **(a)** I only wattmeter is series-parallel
50. **(c)** effective difference DEF 100

## ANSWERS JOURNEYMAN CLOSED BOOK EXAM #4

1. **(b)** electrons passing a point
2. **(a)** series
3. **(a)** one coil
4. **(b)** ammeter
5. **(c)** grounded T. 110-26a condition 2
6. **(c)** lighting
7. **(c)** increases the resistance
8. **(d)** effective value
9. **(b)** 1-6 2-5 3-4-7
10. **(c)** 36" edge of basin 210-52d
11. **(b)** 2 hot wires use neutral
12. **(b)** 75% 220-17
13. **(c)** I & II PVC or bakelite
14. **(c)** AC and DC tungsten 380-14b
15. **(b)** fuse DEF 100 over 600v
16. **(b)** service-ent conductors DEF 100
17. **(d)** Article 480
18. **(d)** I, II, & III chain wrench
19. **(c)** hacksaw and ream
20. **(d)** 50 pounds fixture 410-16a
21. **(a)** yes 300-3c1
22. **(b)** local Code when more stringent
23. **(b)** VD is a percentage
24. **(c)** insulation 310-10
25. **(d)** zinc finish
26. **(b)** saber saw
27. **(a)** 6-32 x 1"
28. **(c)** be alert at all times
29. **(a)** 90 degrees
30. **(a)** 3Ω will consume the most power
31. **(a)** 35 pounds ceiling fans 422-18
32. **(d)** use a chalk line
33. **(c)** silver improves continuity
34. **(c)** perform their duties properly
35. **(d)** level
36. **(a)** hardened steel surface
37. **(c)** 15 feet over driveways 230-24b
38. **(d)** 60% nipple fill Chapter 9 note 4
39. **(b)** tested to withstand high-voltage
40. **(b)** twisted together tightly
41. **(d)** 12Ω will consume most power in series
42. **(c)** Article 250
43. **(a)** 27 5/16" total sum
44. **(c)** fusestat has different size threads
45. **(c)** symbol for ceiling outlet
46. **(d)** check circuit for a problem
47. **(b)** carborundum
48. **(b)** 0.1875 is the decimal eqivalent of 3/16"
49. **(c)** too much pressure on the drill bit
50. **(b)** L2 fuse is blown

## ANSWERS JOURNEYMAN CLOSED BOOK EXAM #5

1. **(d)** prevent loosening
2. **(c)** saw & ream ends
3. **(b)** voltmeter
4. **(b)** two-gang switch
5. **(d)** LB or T
6. **(d)** housekeeping DEF 100
7. **(d)** I,II, or III DEF 100
8. **(d)** copper 110-5
9. **(c)** voltage drop
10. **(b)** current
11. **(c)** direct current
12. **(c)** piezoelectricity
13. **(c)** expansion joints
14. **(a)** one-half cycle
15. **(d)** I, II & III
16. **(b)** Chapter 5
17. **(b)** are sure the power is turned off
18. **(a)** real power
19. **(a)** accessible 250-68
20. **(d)** capacitance exceeds inductance
21. **(b)** shall 90-5
22. **(c)** longevity 110-3a
23. **(a)** current transformer
24. **(b)** short-circuited
25. **(b)** AC
26. **(c)** 220 $W = E \times I$
27. **(d)** increases as length of wire increases
28. **(c)** safety
29. **(a)** BX
30. **(a)** loose connection
31. **(d)** I, II & III 250-119
32. **(c)** both
33. **(d)** continuously DEF 100
34. **(d)** operation DEF 100
35. **(c)** both DEF 100
36. **(a)** cabinet DEF 100
37. **(c)** 6000 $W = I^2 \times R$
38. **(b)** will not
39. **(d)** larger in total diameter
40. **(b)** 3ø 4-wire
41. **(d)** apply solder to each strand
42. **(d)** make wire pulling easier
43. **(c)** decrease nicking of wire
44. **(c)** oil
45. **(d)** for grounds on 120v circuits
46. **(c)** make a good electrical connection
47. **(a)** filament seldom burns out
48. **(c)** condenser
49. **(d)** ungrounded conductor 240-20a
50. **(b)** festoon 225-6b

# ANSWERS JOURNEYMAN CLOSED BOOK EXAM #6

1. **(d)** temperature
2. **(b)** transformer
3. **(d)** I,II or III 410-16c
4. **(d)** paper
5. **(b)** AWG or CM 110-6
6. **(c)** tubular
7. **(c)** cool & insulate transformer
8. **(c)** carbon
9. **(b)** cover keep person warm
10. **(a)** stop button
11. **(c)** water & apply vaseline
12. **(d)** squirrel cage
13. **(a)** 15 & 20 210-7a
14. **(d)** I, II & III 300-20a
15. **(d)** rectifier
16. **(d)** magnetic effect
17. **(d)** conductance
18. **(d)** I, II & III DEF 100
19. **(d)** mechanical function DEF 100
20. **(d)** carries the unbalanced 310-15b4a
21. **(d)** stationary 550-2 DEF
22. **(b)** free of shorts & grounds 110-7
23. **(d)** noncorrosive
24. **(b)** II only DEF 100
25. **(c)** improve finish of threads
26. **(d)** specific gravity
27. **(c)** temperature
28. **(d)** I,II or III 110-13a
29. **(a)** commutator bar separators
30. **(c)** insufficient pressure at fuse clips
31. **(d)** elect. & mechanically interlocked
32. **(c)** avoid excessive starting current
33. **(c)** motor starter
34. **(c)** burn more brightly
35. **(d)** broken
36. **(d)** either vacuum or gas
37. **(d)** either I or II 230-70a
38. **(d)** all of these
39. **(a)** two 3-way & one 4-way
40. **(a)** artificial respiration
41. **(a)** box end wrench
42. **(b)** ammeter
43. **(d)** csa
44. **(c)** single-pole, double-throw
45. **(c)** resistor
46. **(c)** iron losses
47. **(d)** jerk quickly break any arc
48. **(d)** sustained overload
49. **(d)** NFPA
50. **(b)** LB conduit body

## ANSWERS JOURNEYMAN CLOSED BOOK EXAM #7

1. **(c)** does not absorb much moisture
2. **(c)** can be recharged
3. **(d)** 12' steel tape
4. **(b)** turns-ratio
5. **(b)** csa of the wire
6. **(b)** I only
7. **(b)** use at lower rated voltage
8. **(a)** I only
9. **(c)** compensate for voltage drop
10. **(d)** 1,000,000 ohms
11. **(c)** nicks in the wire
12. **(a)** iron
13. **(a)** protect against shock
14. **(c)** protect the rubber tape
15. **(d)** stepping on a nail
16. **(c)** volt-amps
17. **(d)** 32 teeth per inch
18. **(d)** I, II and III DEF 100
19. **(c)** 30 240-51a
20. **(a)** tube saw
21. **(b)** the user may be injured
22. **(d)** lumens
23. **(b)** 2000
24. **(d)** atom negative charge
25. **(d)** I,II or III
26. **(a)** both I & II 90-4
27. **(c)** hickey
28. **(c)** star drill
29. **(c)** steel wire
30. **(a)** dry stick or dry rope
31. **(c)** .001"
32. **(b)** frequency
33. **(c)** voltage
34. **(d)** switch
35. **(d)** storage batteries
36. **(d)** I,II & III earth resistance
37. **(d)** cut internal threads
38. **(b)** forward stroke only
39. **(a)** switches 1 and 3
40. **(d)** threads per inch
41. **(d)** rawl plugs
42. **(c)** secondary
43. **(b)** rosin
44. **(c)** DC amperes
45. **(c)** electro chemistry
46. **(c)** II and III only
47. **(c)** bell & battery set
48. **(c)** series-parallel wattmeter
49. **(c)** current
50. **(c)** split duplex

## ANSWERS JOURNEYMAN CLOSED BOOK EXAM #8

1. **(c)** 6 volt series-parallel
2. **(d)** locknut outside, bushing inside
3. **(a)** grounded 200-1
4. **(d)** all of these
5. **(b)** becomes stronger
6. **(c)** resistance
7. **(c)** temperature surrounding
8. **(c)** avoid snagging or pulling
9. **(b)** 120v
10. **(a)** remove the fuses
11. **(a)** defective tools cause accidents
12. **(b)** insulation to deteriorate
13. **(b)** even spacing, numerous lights
14. **(b)** accessible
15. **(c)** tungsten
16. **(d)** all of these DEF 100
17. **(c)** 1/2 the R of one conductor
18. **(b)** same
19. **(c)** limit switch
20. **(a)** common magnetic circuit
21. **(d)** current
22. **(a)** DC motor
23. **(c)** stationary portion
24. **(a)** slow down rust
25. **(c)** oil
26. **(b)** to keep surfaces clean
27. **(a)** weatherproof DEF 100
28. **(b)** direct
29. **(a)** likelihood of arcing
30. **(a)** 30 hertz
31. **(d)** watthour meter
32. **(b)** join wires and insulate the joint
33. **(a)** steel
34. **(d)** test lighting circuit for a ground
35. **(a)** use plenty of solder
36. **(b)** the resistance
37. **(b)** locknuts and bushings
38. **(c)** **not** a safe practice
39. **(b)** heat
40. **(c)** connected in one line only
41. **(b)** circuit breaker
42. **(c)** I and IV
43. **(d)** 50Ω
44. **(b)** corrosive
45. **(b)** fuse clips would become warm
46. **(a)** minimum loads 220-3b
47. **(d)** THHN T.310-13
48. **(a)** reamed
49. **(b)** expansion bolts
50. **(d)** fine sandpaper

## ANSWERS JOURNEYMAN CLOSED BOOK EXAM #9

1. **(b)** 3-4wy & 2-3wy
2. **(c)** ease of variation
3. **(b)** copper wire
4. **(d)** electrolyte
5. **(a)** black,red,white
6. **(b)** FPN 90-5
7. **(a)** reduce shock
8. **(d)** all of these
9. **(a)** currents would circulate
10. **(c)** derating of ampacity
11. **(b)** condenser
12. **(b)** feeder DEF 100
13. **(d)** cond. will not turn off
14. **(d)** an impossibility
15. **(c)** hydrometer
16. **(c)** windings are common
17. **(a)** temperature
18. **(a)** tighten the clips
19. **(c)** carbon
20. **(c)** higher volt. & lower current
21. **(b)** copper good conductor
22. **(a)** Kirchoff's law
23. **(c)** pressure
24. **(d)** all of the above 210-21a
25. **(c)** protect from damage
26. **(d)** poor contact
27. **(b)** nuts removed frequently
28. **(a)** sum of individual resistances
29. **(c)** shorter life of bulb
30. **(c)** eddy current loss
31. **(b)** green or green with yellow stripes
32. **(b)** do not wear out as quickly
33. **(d)** orange 215-8 230-56 384-3e
34. **(b)** makes pulling too difficult
35. **(c)** exposed
36. **(d)** electromagnet
37. **(a)** personal injury
38. **(b)** nylon string
39. **(d)** the contact resistance
40. **(c)** relationship between E, I and R
41. **(d)** 3/4" per foot 346-8 345-8
42. **(b)** power factor
43. **(a)** 80% 384-16d
44. **(a)** 5a I = E/R 600/120 = 5
45. **(a)** may conceal weak spots
46. **(a)** CO2
47. **(a)** dry DEF 100
48. **(a)** 1.5
49. **(d)** branch DEF 100
50. **(c)** 24 volts

## ANSWERS JOURNEYMAN CLOSED BOOK EXAM #10

1. **(d)** fused 240-20 380-2b
2. **(d)** I & III only
3. **(a)** 6 feet 210-52a1
4. **(c)** used with other
5. **(d)** 80 % 210-23a
6. **(a)** askarel DEF 100
7. **(c)** fitting DEF 100
8. **(d)** many layers set apart
9. **(c)** can be "shaped" better
10. **(d)** salt water
11. **(c)** not closed DEF 100
12. **(d)** I thru V 90-1b
13. **(c)** phase
14. **(b)** two-wires between 3-way
15. **(b)** size
16. **(a)** permanent air space
17. **(a)** dry chemical
18. **(d)** heat sensing element DEF 100 FPN
19. **(b)** stores
20. **(d)** an explosion
21. **(d)** change DC to AC
22. **(a)** rigidly supported
23. **(b)** handy box
24. **(a)** prevent the frame
25. **(b)** galvanic 346-3a
26. **(d)** controller DEF 100
27. **(c)** pink flamingo
28. **(d)** 42 384-15
29. **(d)** renewable fuse
30. **(b)** strengthening a splice
31. **(d)** all of these
32. **(d)** equal to aluminum T. 310-16
33. **(b)** high resistance
34. **(b)** 2.5 ohm $10\Omega/4 = 2.5$
35. **(b)** plumb bob
36. **(c)** firestopped 300-21
37. **(b)** direct current
38. **(c)** use a template
39. **(c)** less reaming is required
40. **(b)** bushing 370-17b
41. **(d)** reverse any two of the three leads
42. **(b)** 746 watts
43. **(d)** 180va 220-3b9
44. **(c)** AC only
45. **(b)** thermoplastic-moisture resistant
46. **(d)** 3 hours DEF 100
47. **(a)** two or more 210-3
48. **(b)** secondary
49. **(a)** non-interchangeable
50. **(d)** **ampacity** remains the same

## ANSWERS JOURNEYMAN CLOSED BOOK EXAM 11

1. **(a)** surge arrester 280-2
2. **(b)** covered DEF 100
3. **(a)** unity
4. **(a)** reduce the current 240-11
5. **(d)** hp is the output
6. **(c)** current develops heat
7. **(a)** inductive exceeds capacitive
8. **(a)** Allen head bolt
9. **(b)** automatic DEF 100
10. **(a)** 120 volts $W = E^2/R$
11. **(a)** may loosen the insulating tape
12. **(d)** steel bushing not used
13. **(c)** ungrounded conductor for switch 380-2b
14. **(a)** 0.5625 is the decimal for 9/16"
15. **(b)** the cause of accident
16. **(b)** 420 watts total
17. **(c)** stoppage of breathing
18. **(b)** only the current will change
19. **(d)** solderless connections
20. **(c)** polarized plug
21. **(c)** flush eyes with clean water
22. **(d)** I, II, & III lamps & motors
23. **(c)** III only 3-way switch connection
24. **(b)** 600v or less 490-2
25. **(a)** 0.125 csa of bus bar
26. **(c)** increase VD across the connection
27. **(d)** Δ delta symbol
28. **(b)** hysteresis 300-20 FPN
29. **(b)** I & II only switch
30. **(c)** if one person is hurt
31. **(b)** low resistance in closed position
32. **(b)** 30a receptacle
33. **(d)** I, II, III & IV ground resistance
34. **(d)** cost is less for copper
35. **(c)** six lengths of conduit
36. **(c)** all parts of the circuit not in contact
37. **(d)** hydrometer
38. **(d)** low point
39. **(a)** zinc and copper
40. **(a)** carbon dioxide
41. **(c)** safety switch
42. **(b)** solenoid
43. **(d)** Article 490
44. **(d)** threads over entire length
45. **(a)** stretch the rubber tape
46. **(c)** larger in total diameter
47. **(b)** travel reaches a preset limit
48. **(c)** 10 ohms
49. **(a)** 6" Table 300-5
50. **(d)** 6 pounds 410-15a

## ANSWERS JOURNEYMAN CLOSED BOOK EXAM #12

1. **(b)** megger
2. **(d)** tapered thread
3. **(a)** parallel
4. **(b)** consider circuit hot
5. **(c)** air
6. **(b)** toggle bolts
7. **(b)** less than the low resistance
8. **(d)** powdered soapstone
9. **(d)** wattmeter
10. **(c)** main DEF 100
11. **(d)** lower the resistance
12. **(d)** protect end of wire
13. **(a)** LL conduit body
14. **(a)** the use of flux
15. **(a)** reduce shock hazard
16. **(a)** underwriters'
17. **(c)** general purpose DEF 100
18. **(b)** less than any one resistor
19. **(c)** converts into mechanical
20. **(c)** I and III only
21. **(b)** power factor
22. **(c)** 10 ohm resistor
23. **(b)** 1/120
24. **(c)** ampacity DEF 100
25. **(c)** reaming the ends
26. **(b)** 1000
27. **(c)** grounded 380-2a
28. **(d)** conductivity
29. **(d)** hacksaw and file
30. **(d)** relay
31. **(d)** clamped perpendicular
32. **(b)** shorted
33. **(a)** volt
34. **(c)** wet location
35. **(a)** may transmit shock to user
36. **(a)** impedance
37. **(a)** rotometer
38. **(d)** all of these
39. **(b)** voltmeter, ohmmeter, ammeter
40. **(c)** iron
41. **(b)** AC or DC
42. **(a)** cartridge fuses
43. **(a)** hanging fixture
44. **(d)** an approved box hanger
45. **(b)** ohm
46. **(c)** supplied by transformers & batteries
47. **(b)** rectifier
48. **(d)** three-way
49. **(d)** non-automatic
50. **(b)** parallel

## ANSWERS JOURNEYMAN OPEN BOOK EXAM #1

1. **(c)** #12 copper 230-31b ex.
2. **(d)** 60-50 422-11b
3. **(d)** 12 times 300-34
4. **(c)** 200 amps T.318-7b2
5. **(d)** 2001 volts 326-1
6. **(a)** #16 minimum T. 402-5
7. **(c)** #4/0 310-15b6
8. **(a)** one is required 210-52d
9. **(b)** enamel 250-96
10. **(b)** 3"x 2"x 2 1/4" T.370-16a
11. **(a)** 167% 450-4a ex.
12. **(d)** 200va 220-12
13. **(d)** steel EMT 348-5 ex.
14. **(a)** #4/0 copper 339-1a
15. **(b)** 200% T.430-152 note 3
16. **(a)** 5 feet 680-70
17. **(a)** water pump 230-72a ex.
18. **(a)** damp location DEF 100
19. **(b)** one foot 470-18c
20. **(c)** accessible DEF 100
21. **(a)** AHJ 90-4
22. **(b)** 13 receptacles 605-8c
23. **(c)** 3 hours 450-42
24. **(a)** #18 400-13
25. **(c)** not more than 6" 110-26a3
26. **(c)** bathroom 210-8b1
27. **(b)** 20 ampere 660-9
28. **(d)** drive through door 210-70a2
29. **(d)** interrupting rating DEF 100
30. **(d)** I,II,III & IV 410-14a
31. **(c)** II & III 230-66
32. **(c)** 4" 318-10b
33. **(a)** I only 450-4a
34. **(d)** 10' 362-22
35. **(c)** 12" 450-21a
36. **(d)** 50 volts 460-6a
37. **(b)** II only 225-31
38. **(c)** manufactured phase 455-9
39. **(b)** 30 amps 230-79b
40. **(d)** I,III & IV 680-4
41. **(d)** 800 amps 240-3c
42. **(b)** II only 318-11a1
43. **(d)** tampering 230-93
44. **(a)** luminaire 410-1 FPN
45. **(d)** I,II or III 250-30a1
46. **(c)** equipment bonding 250-146
47. **(d)** 24" 402-9b
48. **(d)** I,II & III 240-60c
49. **(c)** 10' 352- 47a
50. **(b)** 20 amps 430-53a

## ANSWERS JOURNEYMAN OPEN BOOK EXAM #2

1. **(c)** service drop 230-21
2. **(a)** 1/8 hp 422-35
3. **(c)** I or II 240-30a1
4. **(b)** less per AHJ 430-26
5. **(b)** 50 volts 445-6
6. **(c)** III only 450-6
7. **(c)** conspicuous 110-27c
8. **(a)** 8' 225-19a
9. **(c)** III only 318-8e
10. **(b)** receptacles 210-50b
11. **(d)** Listing 110-3b
12. **(d)** not required 250-68 ex.
13. **(a)** #16 680-25b5
14. **(d)** flexible conduit 300-4a2 ex.
15. **(b)** 1ø - 3ø 240-85
16. **(d)** I,II,III or IV 280-21
17. **(a)** I only 370-40d 250-148a
18. **(d)** I,II or III 410-15b 410-16a
19. **(b)** 1/8" 410-50
20. **(d)** grd. electrode cond. 25024a
21. **(a)** 24" 110-26c
22. **(d)** I,II & III 200-10e
23. **(c)** box listed 370-27b
24. **(d)** I,II & III 210-52g
25. **(a)** 2 1/4" x 4" 370-17c ex.
26. **(a)** 50 amps 680-22c
27. **(a)** 150°C 410-65b
28. **(c)** lighting track 410-100
29. **(d)** 150% 430-6c
30. **(c)** cooking unit 422-32 a&b 422-33
31. **(b)** protected sprinkler 450-42 ex.
32. **(a)** 2" 480-6
33. **(b)** 25% T.352-45
34. **(d)** none of these 300-22a
35. **(d)** 75% 352-29
36. **(c)** 30 amps 373-11b
37. **(b)** 1" 354-3b
38. **(d)** bathrooms 680-71
39. **(d)** .581 sq.in. Chapter 9 Table 4
40. **(d)** I,II or III 427-37
41. **(b)** 10' 680-41a
42. **(d)** thermally DEF 100 over 600v
43. **(d)** 100 pounds 110-31c
44. **(c)** "No Equipment Ground" 210-7d3b
45. **(a)** office bldg. 210-8a6 210-8b1&2
46. **(a)** outdoor outlets 210-52b2
47. **(d)** 600va 551-73a
48. **(b)** 12" 600-10c2
49. **(c)** locked open position 424-19b1
50. **(b)** 5' 250-50

## ANSWERS JOURNEYMAN OPEN BOOK EXAM #3

1. **(a)** 15 384-32
2. **(b)** 230-3
3. **(d)** FCC 328-2
4. **(a)** recpt. outlet 210-50a
5. **(c)** available 110-9
6. **(d)** 0.06" 250-52d
7. **(a)** I & II only 427-2 FPN
8. **(d)** 75 362-7
9. **(c)** III only 410-57e
10. **(c)** 30 600-5b2
11. **(b)** 1/3 430-81c
12. **(c)** 8 410-38b
13. **(c)** 36" T. 110-26a
14. **(b)** 194° F 333-20 ex.
15. **(c)** 10' 210-52h
16. **(a)** 3/4" T.384-36
17. **(b)** 20 480-5b
18. **(c)** 1/2" 410-46
19. **(c)** III only 250-178
20. **(a)** reduce 240-11
21. **(a)** 1 va T.220-3a
22. **(c)** 70% 551-71
23. **(b)** 3' 410-4d
24. **(b)** voltage 424-35
25. **(c)** hexagonal 240-50c
26. **(b)** I & II only 410-8b1 & b2
27. **(c)** in the field 424-29
28. **(d)** 150v 240-50a2
29. **(b)** solely by enamel 348-5(2)
30. **(d)** 15a @ 125v 430-42c
31. **(d)** I,II & III 328-11
32. **(d)** 115% 440-12a1
33. **(c)** 90° C 410-68
34. **(d)** 210-52
35. **(d)** 36" 328-10
36. **(c)** overhead spans 225-26
37. **(b)** 100% 220-15
38. **(b)** feeder 430-2
39. **(a)** 6' 430-145b
40. **(d)** 12.5 - #12 - 20a 210-23a
41. **(b)** appliances 338-3c
42. **(c)** at standstill 430-7c
43. **(b)** as low as practicable 460-8b2
44. **(d)** I,II & III 336-21
45. **(d)** I,II & III 680-24
46. **(a)** factory-installed internal 90-7
47. **(b)** 3' 225-19b
48. **(c)** continuous duty T.430-22b note
49. **(a)** ungrounded conductor 240-20a
50. **(a)** distinctive 200-6a1

## ANSWERS JOURNEYMAN OPEN BOOK EXAM #4

1. **(b)** SWD 240-83d
2. **(c)** II & III 250-62
3. **(c)** 115% 430-110a
4. **(d)** 1/2" 410-66a
5. **(b)** 2' 230-54c ex.
6. **(a)** 1 foot 220-3b8b
7. **(a)** 7 pound-inches 430-9c
8. **(d)** white 200-9
9. **(c)** 15w 424-99b
10. **(c)** II & III 210-8a2 ex.2
11. **(b)** I or III 422-61
12. **(d)** 1000a 230-95
13. **(b)** 50' ... 1/3 364-11 ex.
14. **(c)** III & IV Chapter 9 note 4
15. **(c)** freedom from hazard 90-1b
16. **(c)** 50% 440-62c
17. **(c)** 3 overloads T.430-37
18. **(c)** #1 310-4
19. **(a)** sealed 300-7a
20. **(b)** I & III 347-5,6,8
21. **(c)** #12 250-122a
22. **(d)** 3" 384-10
23. **(b)** 4' 680-21a5
24. **(c)** 3va T.220-3a
25. **(c)** 1/4" 370-18
26. **(a)** not required 240-10
27. **(d)** 150v 250-174c
28. **(c)** cable tray 318-2
29. **(c)** by hand 331-1
30. **(b)** 80% T.220-18
31. **(b)** 6' 250-106 FPN2
32. **(b)** 5' 364-5
33. **(c)** omit the smaller 220-21
34. **(a)** metal water pipe 250-50a
35. **(d)** I & III 110-3a2,6
36. **(d)** I,II & III 342-3c 336-5a1
37. **(a)** 8' 363-18
38. **(d)** 3' 410-27c
39. **(d)** I,II & III 365-2a
40. **(c)** flame arrestor 480-9a
41. **(b)** 10° C 310-13 FPN 402-3 FPN
42. **(d)** pendants, lamps, cables 400-8
43. **(d)** 100a 230-79c
44. **(d)** 16" 410-15a
45. **(b)** 5' 410-101c8
46. **(b)** 30 conductors 362-5
47. **(a)** D 310-11c
48. **(d)** I,II or III 210-4d
49. **(b)** II and III only 220-3a
50. **(d)** grounded 410-23

## ANSWERS JOURNEYMAN OPEN BOOK EXAM #5

1. **(b)** will not 230-95 FPN 1
2. **(a)** 300-5e
3. **(b)** 8" 424-39
4. **(b)** 6" 300-14
5. **(c)** I & II 220-17
6. **(c)** stranded type 225-24
7. **(b)** 1 1/2" 410-18a
8. **(a)** dust 500-8
9. **(b)** 3/4" 346-8 345-8
10. **(b)** shall Chapter 9 note 3
11. **(b)** 1/4" 347-9
12. **(b)** 200va 220-12
13. **(b)** #4 370-28a
14. **(b)** 6' 250-102e
15. **(a)** direct sunlight 347-1,2f
16. **(d)** all of these 330-20,22
17. **(c)** I or II 550-5a
18. **(c)** kva 430-7b1
19. **(b)** 4' 422-16b2b
20. **(a)** sub. increased 300-21
21. **(c)** workmanlike 110-12
22. **(a)** 12 linear feet 210-62
23. **(c)** create a hazard 240-3a
24. **(c)** 36 times 370-71b
25. **(d)** 12" 470-3
26. **(b)** first-make, last break 250-124a
27. **(a)** lowest 310-15a2
28. **(d)** do not project 318-8a
29. **(b)** control selected DEF 100 (over 600v)
30. **(a)** #10 110-14a
31. **(b)** 2" 331-5b
32. **(c)** T.402 402-5
33. **(d)** 3 1/2" 410-38c
34. **(b)** fan circuit 424-63
35. **(a)** 5' 680-6a1
36. **(c)** I,II & III 410-24a
37. **(a)** 3' T.110-26a
38. **(c)** 6 pounds 410-15a
39. **(a)** electrode conductor 250-24b1
40. **(d)** I,II or III 339-3a4
41. **(b)** FC 363-1
42. **(a)** 2' 220-12b
43. **(b)** II only 220-30b
44. **(c)** voltage drop 230-31 FPN
45. **(c)** 18' 210-6d1 b
46. **(a)** wet DEF 100
47. **(b)** #6 copper 280-23
48. **(d)** 1.2v 480-2
49. **(a)** approved 110-2
50. **(b)** 6' 250-56

## ANSWERS JOURNEYMAN OPEN BOOK EXAM #6

1. **(a)** adeq. bonding & grd. 250-116 FPN
2. **(b)** #1/0 318-3b1a
3. **(b)** II only 200-2
4. **(b)** support fixtures 347-3b
5. **(c)** #12 410-105a
6. **(b)** yoke 210-4b
7. **(a)** exposed 365-2a
8. **(a)** isolating switches 380-13a
9. **(b)** header duct 356-1 358-2
10. **(b)** hysteresis 300-20 FPN
11. **(d)** 90° C 410-5
12. **(b)** 115% 445-5
13. **(c)** reamed 346-8
14. **(a)** #8 680-22
15. **(a)** 24" 333-7b2
16. **(a)** 0.053 373-10b
17. **(d)** 6' 6" 110-26e
18. **(a)** equal to maximum 300-3c1
19. **(d)** disconnect 230-75
20. **(b)** 6' 210-50c
21. **(d)** small appl. circuit 210-52b2 ex.1
22. **(a)** 1/16" 300-4a1
23. **(a)** 6' 365-6c
24. **(b)** 200a 110-26e ex.
25. **(c)** both I & II 410-30b
26. **(c)** #8 320-8
27. **(b)** Cover T.300-5
28. **(c)** III only T. 400-4
29. **(a)** hazardous location 330-3
30. **(c)** 6' 410-67c
31. **(b)** separate box 370-28d
32. **(c)** 1/2 ohm
33. **(b)** 8' 300-5d
34. **(c)** 8 times 370-28a1
35. **(c)** either I or II 422-16b3
36. **(c)** I,II & III 210-7f
37. **(b)** 70% T.220-20
38. **(c)** not true 1 3/4 kw T.220-19
39. **(d)** I,II & III 230-92
40. **(d)** I,II or III 250-70
41. **(d)** I,II & III 318-5a,b,d
42. **(c)** 45% 310-15b2a
43. **(d)** inductive current 300-20a
44. **(a)** #10 cu 225-6a1
45. **(b)** 12" 336-18
46. **(c)** end seal 310-15b7
47. **(a)** GFCI 680-31
48. **(a)** solder 230-81
49. **(a)** EMT 300-22b
50. **(b)** 3' 230-9 ex.

## ANSWERS JOURNEYMAN OPEN BOOK EXAM #7

1. **(b)** ungrounded 380-2a
2. **(b)** X 90-3
3. **(a)** #4 200-6b 310-12a
4. **(d)** impedance protected 430-7a 14
5. **(c)** grd. terminal ser. equip. 680-25d
6. **(c)** both I & II 110-10
7. **(c)** 86° F T.310-16
8. **(c)** 35kv 450-24
9. **(d)** MI cable 330-15
10. **(c)** 70% 220-22
11. **(c)** NFPA 90-6 FPN
12. **(a)** 14' 600-9a
13. **(a)** I only 430-74b
14. **(a)** surge arrester 280-2
15. **(d)** I,II & III 328-31
16. **(d)** 150v 680-20a2
17. **(b)** suitable 110-8
18. **(c)** lateral DEF 100
19. **(c)** raintight 348-10
20. **(a)** highest 430-7b3
21. **(c)** I,III & IV T.220-3a 210-11c(1)(2)
22. **(b)** Article 225 110-31b1
23. **(c)** both I & II 220-32a1,3
24. **(c)** both I & II 321-4
25. **(a)** 15 & 20 210-7a
26. **(c)** gases or vapors 500-7
27. **(d)** 310-15b5, 250-118, Chapter 9 note #3
28. **(c)** 0.030 380-9 410-56d
29. **(c)** locked 430-102 ex.1
30. **(b)** grounding 250-119
31. **(c)** 4' 424-59 FPN
32. **(d)** 20a 210-23 T.210-24
33. **(a)** inversely T.310-16
34. **(a)** 10' 680-6a1
35. **(b)** temp. limiting 422-13
36. **(a)** supported by messenger 340-4(2)
37. **(c)** 25a 210-21a
38. **(a)** 1/4" 370-20
39. **(c)** raintight to drain 225-22 230-53
40. **(b)** 48" 230-54c ex.
41. **(b)** 0.017 Table 8 310-3
42. **(c)** portable generators 210-7b ex.1
43. **(b)** arms & stems 410-28c
44. **(a)** 18" 210-52a3
45. **(d)** both I & II 230-82(1)(3)
46. **(a)** armored cable 333-3
47. **(b)** 0.5 or larger 220-2b
48. **(c)** 75% 220-17
49. **(a)** 119a T.325-14
50. **(d)** I,II,III & IV 370-23d1

## ANSWERS JOURNEYMAN OPEN BOOK EXAM #8

1. **(a)** 70% 310-15b2a
2. **(b)** #4 300-4f
3. **(c)** back fed 384-16g
4. **(c)** listed for raceway 410-31 ex.1
5. **(d)** 10' 440-64
6. **(a)** 1/8 hp 422-31a
7. **(c)** less than Tables 4 & 5
8. **(c)** pigtail to silver terminal 300-13b
9. **(b)** 8' 6" T.110-34e
10. **(d)** **not** true 210-70a
11. **(c)** both I & II 230-23a
12. **(d)** I,II & III 250-140(1)(2)(3)
13. **(b)** 15a T.210-24
14. **(d)** 8' 230-24a
15. **(c)** 30a 215-2b1
16. **(d)** .8 & larger Chapter 9 Table 1 note 7
17. **(d)** 6' 7" 380-8a
18. **(d)** 3 conductors T.370-16a
19. **(b)** 10' 230-24b
20. **(a)** 7' 424-34
21. **(d)** aquarium 250-114(3)b
22. **(c)** 100' T.300-19a
23. **(d)** 24" T. 300-5
24. **(d)** #4 copper T.250-66
25. **(c)** 1.2426 Table 4 (csa x 60%)
26. **(d)** I,II,III or IV 250-64b
27. **(d)** 20 pounds 325-21
28. **(a)** I or II 338-3a
29. **(c)** both I & II T.400-4 note #5
30. **(b)** enclosed 215-4b
31. **(d)** 8 3/4 kw 210-19c
32. **(d)** 20' 680-6a3
33. **(c)** 80% 210-23a
34. **(b)** for wet locations 410-4a
35. **(c)** nonlinear 220-22
36. **(d)** metal plugs & plates 370-18 373-4
37. **(d)** 5/8" 250-52c2
38. **(b)** high temp. 351-4b2
39. **(b)** #10 545-4b
40. **(d)** 10' 680-51e
41. **(d)** I,II & III 410-16d
42. **(c)** II,III & IV 422-16b1(b)(c)(d)
43. **(a)** 17 1/2" T.349-20a
44. **(c)** ambient temp. 310-10 (4)
45. **(c)** either I or II 333-19
46. **(a)** 90° C T.310-13
47. **(b)** 12" 225-14d
48. **(b)** 50 pounds 410-16a
49. **(c)** 5000a 240-83c
50. **(d)** all of these DEF 100

## ANSWERS JOURNEYMAN OPEN BOOK EXAM #9

1. **(c)** hook sticks 364-12
2. **(a)** .040" 410-38a
3. **(b)** 15v 680-20a1
4. **(b)** 25 kva 450-11
5. **(b)** Coordination 240-12
6. **(d)** grounded neutral 210-10 215-7
7. **(b)** 220-11 220-16
8. **(b)** I or II 225-4
9. **(b)** listed for 250-70
10. **(d)** 30a 328-6b
11. **(a)** enclosed in 410-54a
12. **(b)** 18" 250-64a
13. **(b)** water accumulation 410-57f
14. **(b)** 6" 380-5ex.
15. **(c)** 1 1/4" Tables 4 & 5
16. **(c)** 5 3/4" T.346-10 ex.
17. **(b)** 55 215-2c
18. **(c)** 5 1/2' 210-52
19. **(b)** #12 680-25c
20. **(a)** more than 10% 384-14
21. **(d)** .026 Table 5*
22. **(c)** cable assemblies 250-86 ex.2
23. **(b)** solder 250-70
24. **(c)** 20' 250-50c
25. **(b)** 25a 210-3
26. **(c)** 40% 354-5
27. **(b)** 2 cu.in. T.370-16b
28. **(b)** #3/0 T.250-66
29. **(d)** 10' 331-15
30. **(d)** one cond. diameter 365-3d
31. **(b)** Group-operated 460-24a
32. **(d)** 5' 680-12
33. **(d)** motor-overload device 430-32d
34. **(a)** 1/4" 240-32 373-2a
35. **(d)** all of these 305-4c,h
36. **(d)** 2" 342-7a1
37. **(d)** 120 gallons 422-13
38. **(a)** 1/4" 600-41c
39. **(d)** I,II & III 430-82b
40. **(c)** 18" 250-86 ex.3
41. **(b)** immediately 305-3d
42. **(c)** I & II 326-4
43. **(c)** cover 370-25
44. **(b)** bonding 250-90
45. **(d)** grounded 200-7
46. **(d)** 25' 250-86 ex.1b
47. **(d)** I,II or III 250-56
48. **(c)** 6440 T.430-148 F.L.C. va = E x I
49. **(d)** #10 T.250-122
50. **(d)** 16' T.346-12b2

## ANSWERS JOURNEYMAN OPEN BOOK EXAM #10

1. **(c)** I & III only 210-6a1,2
2. **(c)** I & II 110-11
3. **(c)** 112 1/2 kva 450-21b
4. **(b)** twice 370-16c
5. **(b)** 2' 328-31
6. **(c)** grounded DEF 100
7. **(b)** .109 Table 8
8. **(d)** Wooden 110-13a
9. **(a)** 3" 410-66b
10. **(c)** I,II & III 230-50a
11. **(b)** #12 225-6b
12. **(c)** 601a 240-6 ex.
13. **(c)** cma 250-95
14. **(b)** 300v 328-6a
15. **(b)** II only 110-26b
16. **(a)** adequately bonded 365-2a
17. **(b)** elect. continuous 250-64c
18. **(c)** I or II 215-4a
19. **(a)** varying duty DEF 100
20. **(d)** raceway 300-5c
21. **(c)** 1/4" 370-17c
22. **(d)** 1000a 374-6
23. **(a)** 40% T.220-30
24. **(b)** 5' 680-22a5
25. **(a)** energized 300-31
26. **(d)** II,III & IV DEF 100
27. **(c)** MC 334-1
28. **(a)** round 370-2
29. **(d)** unswitched 410-6
30. **(c)** good continuity 250-12
31. **(c)** 12" T. 300-5
32. **(a)** 60° C 336-26
33. **(d)** removed from raceway 356-7
34. **(a)** 1 1/4" 300-4a1
35. **(b)** galvanic action 345-3a 346-3a
36. **(a)** 5 times 330-13(1)
37. **(a)** 410a 220-22
38. **(b)** 12' 230-24b
39. **(b)** MI 200-6a1
40. **(c)** 48a 422-11f1
41. **(c)** #14 copper T.310-5
42. **(d)** fibers or flyings 500-9
43. **(b)** four 348-12
44. **(a)** 75% 352-7
45. **(b)** 8000va 220-12a
46. **(b)** destructive corrosive 334-4
47. **(d)** seal 330-15
48. **(a)** 310-15b2 ex.3
49. **(d)** Table 8
50. **(c)** Table 5a

## ANSWERS JOURNEYMAN OPEN BOOK EXAM #11

1. **(b)** branch-circuit 210-20a
2. **(a)** bare copper 338-1b
3. **(c)** operates 110-4
4. **(c)** II & III 230-6
5. **(c)** 135% 460-8a
6. **(b)** serious degradation 310-10 FPN
7. **(d)** all of these 250-30a2 ex.
8. **(d)** I,II & III 210-25
9. **(d)** 100' 240-4b2
10. **(c)** both I & II 230-71b
11. **(d)** 60° C 339-5
12. **(a)** individual O.C.P. 410-103
13. **(a)** 2.071 Table 4
14. **(b)** 300w 422-48b
15. **(a)** .0625" 370-40b
16. **(c)** receptacle listed 410-56f3
17. **(c)** 15a 210-3
18. **(c)** simult. 380-2b ex.1
19. **(b)** 300va 422-31a
20. **(b)** 50a T.310-16 & T.310-13
21. **(c)** 4 1/2' 350-18
22. **(c)** multi. assembly DEF 100 353-1
23. **(c)** right angle 358-5
24. **(c)** both a & b 354-6
25. **(d)** I,II,III or IV 427-36
26. **(a)** 6" T. 300-5
27. **(b)** 1/8" 370-21
28. **(c)** II & III 324-3 (1), (2)
29. **(b)** complete 300-18
30. **(a)** 50w 422-43a
31. **(d)** 25 ohm 250-56
32. **(b)** .0353 Table 5
33. **(d)** I,II & III 210-8a5
34. **(c)** I & III 110-21
35. **(d)** all of these 250-104a1
36. **(c)** 18" T.300-5
37. **(d)** need not be polarized 328-20
38. **(d)** I & II 333-3 333-4
39. **(a)** #1/0 356-4
40. **(c)** I or II 250-94 (2)(3)
41. **(d)** I,II,III & IV 310-10
42. **(d)** 50% 336-5a1
43. **(a)** not be burned 240-41a
44. **(b)** structural ceiling 384-4
45. **(b)** interrupting 110-9
46. **(c)** 660 ... 750 210-21a
47. **(b)** 125% 215-2a
48. **(c)** III only 250-2d
49. **(c)** NMC 336-30a2
50. **(c)** clothes closets 240-24d

## ANSWERS JOURNEYMAN OPEN BOOK EXAM #12

1. **(b)** .0209 Table 5*
2. **(d)** voltage drop 310-15 FPN
3. **(d)** I,II & III 410-29c
4. **(b)** I & II 422-61
5. **(c)** panelboard 210-4a
6. **(d)** I,II & III 310-4
7. **(c)** 10 penny 320-7
8. **(c)** total amp rating 220-4b
9. **(a)** 10' 225-4
10. **(c)** grounded 200-3
11. **(b)** equivalent to 110-14b
12. **(b)** 50% 210-23a
13. **(c)** dry locations 353-2a
14. **(c)** both a & b 210-4c ex.1,2
15. **(a)** watts 430-7
16. **(c)** insulation 310-10
17. **(b)** braces or guys 230-28
18. **(d)** I,II & III 110-33b
19. **(d)** basin DEF 100
20. **(d)** I,II or III 225-12
21. **(b)** fished in voids 336-4a
22. **(c)** #2/0 310-15b6
23. **(b)** bathrooms 240-24e
24. **(d)** I,III & IV 424-19
25. **(b)** not be less than 230-23c 250-24b1
26. **(b)** shall not prohibit 430-14b ex.
27. **(c)** 42 devices 384-15
28. **(c)** 1000v 250-24b
29. **(d)** I,II & III 380-14a1,2,3
30. **(c)** carpet squares 328-1
31. **(b)** 2 outlets 210-70a
32. **(b)** beginning of installation 215-5
33. **(c)** 6' 333-12a
34. **(c)** number plus one 680-21d
35. **(c)** both I & II 230-7 ex.1 & 2
36. **(a)** 6" 250-92 FPN
37. **(c)** either I or II 280-11
38. **(d)** shall not exceed 60 days 305-3b
39. **(d)** 2' 310-15b2a
40. **(d)** 500 sq.ft. 220-3c1
41. **(b)** 6' 225-19e
42. **(c)** 120/240v Appendix D Examples
43. **(b)** carry the unbalance 310-15b4a
44. **(c)** III only 430-71 DEF
45. **(d)** removed 354-7
46. **(d)** 40' 225-6b
47. **(c)** 100% 210-21b1
48. **(b)** 30' 374-2
49. **(c)** shielded 310-6
50. **(c)** CM 110-6 310-11a4

## ANSWERS JOURNEYMAN OPEN BOOK EXAM #13

1. **(d)** 2 1/2" T.230-51c
2. **(b)** II only 430-12a
3. **(a)** mogul 410-53
4. **(c)** 15/16" 370-24
5. **(b)** garage 210-8a2
6. **(c)** I,II & IV 220-3a
7. **(a)** optional method 220-30a
8. **(c)** air ducts 250-104c FPN
9. **(a)** I or III 424-42
10. **(d)** temperature rise 364-23
11. **(d)** #10 T.350-12
12. **(a)** 50% 380-14b2
13. **(a)** 6' 210-52a
14. **(d)** Stranded 410-28e
15. **(b)** 10' 680-8(1)
16. **(c)** govern. bodies 90-4
17. **(c)** 70% 220-22
18. **(d)** 200a 384-16c
19. **(b)** blue 424-35
20. **(b)** II or III 410-82
21. **(c)** I or II 422-16a
22. **(d)** I,II & III 225-6b
23. **(d)** all of these 400-10 FPN
24. **(a)** raintight 230-54a
25. **(d)** I or III 230-22
26. **(d)** all of these Chapter 9 note 4
27. **(d)** front and back 210-52e
28. **(a)** 6000a 240-60b 240-6
29. **(d)** none of these 362-9
30. **(c)** copper 110-5
31. **(b)** #18 402-6 410-24 FPN
32. **(c)** 2 sq.ft. 250-83d
33. **(b)** 20a 411-6
34. **(d)** I,II & III 250-2d
35. **(b)** temporary lighting 305-4d
36. **(b)** one-half 320-7
37. **(d)** I,II or III 328-30
38. **(c)** II or III 354-15
39. **(a)** one conductor 370-16b1
40. **(c)** 18 Tables 4 & 5
41. **(c)** 2 1/2' 230-51a
42. **(c)** bonding jumper 250-98
43. **(b)** #8 310-3
44. **(a)** two 20a 210-11c
45. **(d)** not at all 370-20
46. **(c)** 18" 680-20a3
47. **(a)** type AC 333-3
48. **(a)** I only 250-64b
49. **(c)** 75a 240-6
50. **(d)** I,II & III 430-109a1

## ANSWERS JOURNEYMAN OPEN BOOK EXAM #14

1. **(c)** grounded 240-23
2. **(b)** 700a 374-6a
3. **(c)** direct burial 339-3a
4. **(b)** joined mech. 110-14b
5. **(c)** #1/0 358-10
6. **(a)** 300v 352-1b2 352-22b3
7. **(a)** 65a 310-15b2 ex.3
8. **(c)** bushing 300-5h
9. **(d)** I & III only 305-3c
10. **(d)** I,II or III 328-33
11. **(a)** 300v 410-78
12. **(d)** Skin effect heating 426-2
13. **(d)** open or closed 230-77
14. **(c)** 65% T.220-30
15. **(c)** 5kw 220-18
16. **(d)** 90° C T.310-13 410-31
17. **(c)** #18 or larger 240-4b2
18. **(d)** accord. T.250-66 250-102c
19. **(b)** 50v 110-27a
20. **(a)** #10 copper 250-140(2)
21. **(d)** all of the above 90-1c & 90-2c
22. **(c)** both I and II 210-8a1 & 210-52d
23. **(c)** 6530cm Table 8
24. **(a)** 8 breakers 384-15
25. **(b)** 143a T.430-150
26. **(d)** colored stripe 210-5a 200-6d
27. **(c)** 25' 240-21b2a
28. **(c)** NEC Tables 430-6a 430-52
29. **(d)** 1/2" to 4" 348-7a,b
30. **(c)** both I & II 225-25
31. **(c)** both I & II 300-17
32. **(d)** 35,000v 326-3
33. **(b)** II or III 240-41b
34. **(c)** 3.5 kva Appendix D example D2b
35. **(d)** 8' 250-52c3
36. **(d)** I,II & III 310-4 FPN
37. **(a)** receptacle 328-14
38. **(b)** 3/8" 410-30a
39. **(d)** not required to be access. 250-68a ex.
40. **(a)** dangerous overheating DEF 100
41. **(c)** 10 cu.in. T.370-16a
42. **(c)** 36 1/2' 550-5d
43. **(c)** both I & II 336-6b
44. **(d)** 18' 230-24b
45. **(d)** replacement for existing 240-51b
46. **(a)** 12" 210-52c3
47. **(d)** 14' intervals T. 346-12b2
48. **(c)** orange 215-8 230-56 384-3e
49. **(b)** outside 220-3a
50. **(b)** 1/4" 300-6c